ONE WOMAN AND HER DOG

Viv Billingham

 Patrick Stephens, Cambridge

Dedicated to the memory of Meg (Stud Book No 24202) and to her master GWB who, by allowing me to pick his brains, taught me to use mine.

First published in 1984

British Library Cataloguing in Publication Data

Billingham, Viv
  One woman and her dog.
  1. Sheep dogs—Scotland
  I. Title
  636. 7'3   SF428.6

ISBN 0-85059-704-8

Photoset in 11 on 12 pt Plantin by Manuset Limited, Baldock, Herts. Printed in Great Britain on 115 gsm Fineblade coated cartridge by St Edmundsbury Press, Bury St Edmunds, Suffolk, and bound by Hunter & Foulis, Edinburgh, for the publishers, Patrick Stephens Limited, Bar Hill, Cambridge, CB3 8EL, England.

Contents

Preface

by Major-General D.L. Lloyd Owen, CB, DSO, OBE, MC

When Viv Billingham first told me—rather diffidently, I seem to remember—that she had written a record of her life for the benefit of her young son, Geoff, in future years, I persuaded her to let me read it. That was nearly two years ago. New chapters have been added since then and much else has been enhanced, but it was clear to me from the start that I had read the testimony of someone with a rare insight into sheep and sheepdogs in particular, and an astute understanding of nature and the beauty of our countryside in general. All this was before the addition of a superb collection of photographs. They cannot fail to bring to life some of the human characters and the dogs so admirably described in this attractive book. And her husband, Geoff, known throughout these pages as GWB, had not even begun to create the really exquisite vignettes which adorn the heads of each chapter. I already knew what a fine artist he is for he had given me prints of two of his delightful drawings some years previously.

GWB, like his wife Viv, is a top-class handler of sheepdogs at International level. To be able to portray, as he has done so brilliantly, the flowers and creatures of nature with never a lesson of any sort, shows his talent, his acute powers of observation and his innate sympathy with them. How fortunate Viv is to have these qualities of understanding, too. They shine throughout from the pages of this enchanting chronicle with obvious sincerity.

For the professional shepherd and the sheepdog handler there will be much of interest to think about and distil in this most informative book. For the amateur enthusiast—such as myself—there is a vast fund of perceptive and sound advice, based on firmly held opinions formed during years of practical experience. Written, as this book is, modestly and with a wealth of common sense, few people can fail to learn something from almost every page.

For those readers who have little or no knowledge of dogs and sheepdog trials nor of sheep and the countryside, this book will surely bring much joy. It is full of the simple philosophy of a woman writing about the animals she loves so much, and about the life of a shepherdess which she had to learn the hard way.

Those who read and enjoy this book will not find it hard to understand why 8-year-old Garry, born of GWB's splendid bitch Jed, is the dog which has meant such a lot to Viv Billingham.

David Lloyd Owen
Balintore, October 1983

Foreword

by Eric Halsall

Gael has put a black-throated diver into the cold, clear waters of the lochan, and the little collie flips her tongue with casual interest over the two brown eggs the bird has left on the crude nest of sphagnum moss. I sit down in the heather to watch—and to reflect on my good fortune and privilege that such simple pleasures are everyday happenings. Gael and I are on the Isle of Mull; we share events akin to these at home above the Gorge of Cliviger in Lancashire.

Viv Billingham, the author of this book, enjoys similar incidents among the wild grandeur of the Cheviot hills, and her writings reflect the joy of life in wild places. That alone is sufficient reason for me to commend her writings to all who find pleasure in Britain's unspoiled countryside. Viv endeared herself to five million people who saw and heard her almost poetic descriptions of her Cheviot home on the BBC2 television programme of *One Man and His Dog*. Her book complements those first impressions so that all may share the joys—and tribulations—of life in the Cheviots. Viv is a shepherdess of no mean calibre, and her story is consequently very practical.

Viv assists her husband Geoff (whose sketches add to the fascination of the book) in the management of one of the oldest flocks of South Country Cheviot sheep over 800 acres in the Scottish Borders, and she can lamb a ewe, clip and dip a flock, gather the hillside of sheep, as well as carry out the normal feminine duties of wife, and mother to Geoffrey Junior. She is regarded as an equal in the competitive world of sheepdog trials—quite a feat in what is regarded almost solely as a man's world. Viv is one of very few ladies ever to represent their country in International trials competition. Viv certainly knows the collie dogs which are her constant companions and how to manage her sheep. She loves her dogs, with sentiment, and in a practical, common sense way. She has great insight into the mind and capabilities of the working collie.

This book is written with wit and charm, with feeling and knowledge by a lady who knows what she is writing about, and there is always the spirit and freshness of the wild Cheviot in the words—plus the traditional ways of her Welsh beginnings. At times controversial—which adds so much to the current literature on working collies—the writing is detailed in discussing canine temperament and character and the breeding and care of puppies. Nowhere could you get more help and sound advice for training dogs for work on the hill and in the more precise trials field.

Follow Viv's methods—and reasons—and you will have a collie dog capable of shepherding stock. Follow her advice and you will have a trusted and loyal companion. Read her words and you will enjoy watching sheepdog trials for evermore. And the beauty of wild places will be with you throughout.

Eric Halsall
Cliviger, October 1983

'A working sheepdog must be moulded with great care over a period of time, depending on the quality of the material provided and the intricacy of the shape required until, finally, you behold the finished article, your creation, hopefully a combination of brains, ability and great beauty.' Garry and Laddie (foreground), father and son, in the Cheviots.

Introduction

There are those who may consider a woman presumptuous in putting pen to paper on the hallowed subject of sheepdog training as there are many men more qualified to do so. All I have to say in my defence is that I do not consider myself an expert, or a literary genius. I have either read or acquired most of the books written about sheepdogs and have enjoyed them. However, I feel that there is so much left unsaid on the subject, particularly as the Border collie is rapidly becoming used all over the world.

Over the years, my husband, Geoff, and I have come to the conclusion that there are very few handlers who are willing to disclose their methods of sheepdog training, preferring to keep them a secret. I disclose ours in the genuine hope that 'man's best friend' will be the main benefactor.

It has often been said that there is a man for the dog and a dog for the man. Only on rare occasions are we fortunate in meeting a trainer who is capable of adapting himself to the whims and ways of any dog—a true handler in every sense of the word. Strange as it may seem, the most successful partnerships are often due to similarities in temperament between dog and handler, and occasionally in looks! Although some dogs are impatient and ill-tempered by nature, an impressionable young dog will rapidly become so in the wrong hands.

The following pages are written for people like me, who are learning by experience, and for those who find themselves owners of dogs which are sensitive. (This sort of dog often takes offence when it is necessary to reprimand it and 'sulking' is a common fault in many of the dogs being bred today.)

My dogs have all taught me much more than I could ever teach them as there is a reason for everything which a dog does. When he does make a mistake, it is often because he misunderstood some exercise that the handler has practised with him previously. All animals are creatures of habit and therefore it is important to correct bad habits in a kind way as a dog's life expectancy (compared to a human's) is, after all, relatively short and therefore should be sweet!

In the following pages you may find some repetition. Patience, kindness and common sense cannot be over-emphasised in the creation of a trustworthy and useful partner.

Vivien M. Billingham
Yetholm, October 1983

Chapter 1

Tribute to William Goodfellow and Laddie

William Goodfellow—or Willie as he was known to all of his friends, and rivals—was born in the latter part of the last century. From the tender age of ten, except when he fought in the First World War, Willie was never without the companionship of a dog. After returning unscathed from the war he spent the greater part of his life shepherding the rugged fells of Cumbria and Northumberland. As an assistant shepherd, caring for 500 ewes at Bewcastle, he was paid the then princely sum of £5 per month!

One of his greatest pals was Dave Dickson of Hawick (by all accounts a priceless character), the owner of Ben. It is over 20 years since this highly regarded dog made his mark in the world of sheepdogs. He was known to be a great worker of sheep and represented Scotland on many occasions. Ben sired many highly thought of registered pups and, if the truth were told, numerous highly thought of unregistered ones!

It was well known locally that Dave Dickson enjoyed a dram with his friends. During these social calls Ben was left out of doors to amuse himself. This amusement consisted of assignations of a flirtatious nature with bitches belonging to the local shepherds.

Below left *Garry with sons Laddie (left) and Glen (centre).*

Right *Willie and Laddie* (The Farming News, Glasgow).

Below *Willie presented with shield at Kelso Scottish National* (R. Clapperton, Selkirk).

Ben was praised for his great intelligence and good temper which, fortunately, the larger part of his progeny inherited from him. Bare-skinned, medium-sized and more black than white, one of his best skills was his ability to cut out and hold a single sheep away from the rest of the flock. His sire, Crozier's Toss, was also particularly good at shedding, and his dam, Robertson's Beat was known to be a sound worker on the hill.

Willie Goodfellow always showed a genuine affection towards all dogs, and counted himself very fortunate in being able to obtain one of Ben's sons out of Phil—Laddie was his name. He was a slightly reticent, medium-coated, tri-coloured dog who, according to his master, was an even better dog than his illustrious sire and who went on to do great things for Willie, winning and gaining places at numerous sheepdog trials throughout the region.

Willie and Laddie's proudest moment came in 1958 when Laddie was 8 years old—they won the championship at the Kelso Scottish National Sheep Dog Trials. Up to that date, no Sassenach had won this great honour.

Just like his sire, Ben, and grandsire, Toss, Laddie proved himself to be a powerful worker and a particularly good shedding dog. He was born with a natural ability to line sheep up on his own and keep them on course with few, or no, commands.

Willie was our close friend for over 20 years. He was one of the finest men you could ever wish to meet—a great storyteller and humorist, despite the fact that he was often in pain as he had had both of his legs amputated when he was in his 70s. Undaunted by this tragedy, not only did Willie learn to walk on artificial

Below left *Willie and Laddie (bottom right), Scottish Team, Kelso* (R. Clapperton, Selkirk). **Above** *Jimmy Anderson's dog Hope.*

legs, but he also continued to handle a dog and took on the occasional judging stint as well.

One of Willie's nicest features was that he showed immense interest in other people's dogs. On his annual winter visit to our home, he could not settle in the house until he had inspected the contents of each kennel and, afterwards, insisted that we place a few sheep at a convenient vantage point so that he could assess the 'shape' and ability of each young dog in turn. He was always willing to give his much-valued kindly and practical advice.

After tea on these visits, as we sat warming ourselves by the fireside, the conversation would somehow get around to Laddie's exploits. A light would shine in Willie's clear, grey eyes as he reminisced about the days when men were men and dogs were dogs and Laddie was in his heyday. In those days, sheepdog trials were occasions when everyone got together to have fun, relax and enjoy themselves generally.

Willie spent his latter working years shepherding the grass parks of the Netherby Estate, sometimes caring for a thousand sheep. After he retired he lived in the cottage at Kirk Andrew Towers until his death at the age of 82 in 1980, eight days after the death of his old friend and rival, Jimmy Anderson of Jedburgh, whose famous dog, Hope, he had much admired.

The priceless characters of the trial scene, who are no longer with us, can never be replaced, and therefore must never be forgotten. Greatly loved and respected by all who were fortunate enough to know him, Willie is remembered mainly for his great courage and his many achievements with his favourite collie, Laddie. I owe him so much and here pay tribute to him for the inspiration which he has given to me.

Chapter 2

Early memories

When I was a small child about 4 years old, I accompanied my mother and younger sister to a sheepdog trial, my very first! It was held in a field on the outskirts of Norton, not far from the 'duck pond' where we regularly picnicked on warm summer days. I remember vividly my first sighting of a real shepherd and his dog and can still recall how enraptured I felt by the entire proceedings. For me it was definitely a case of 'once seen never forgotten'; I was well and truly smitten.

From that day on my ambition was to be a shepherdess. (I was so impatient to grow up, and desperate to learn how to train a sheepdog.) As well as yearning to become a shepherdess, I confess to further ambitions, all in some way closely connected. One ambition was quite small and therefore not too difficult to achieve, the other, for a girl, I considered quite an achievement, if I could master the art.

Believe it or not but my smallest ambition, as a 7-year-old, was to pick up and cuddle a live lamb! Although I had seen plenty skipping and playing together along the banks and hedges near my home, I had never actually touched one, never felt its soft woolliness against my cheek, and I was very anxious to do just

that. One quiet Sunday afternoon, whilst out for a walk, my big chance finally arrived for, under a shady hawthorn, sleeping among the last of the spring daffodils, lay a large black woolly lamb.

Inch by inch I crept up quietly and, with bated breath, I carefully picked it up. My disenchantment will never be forgotten. Lambs, like everything else, are not what they seem. This one, instead of being soft and woolly to the touch, had a harsh, coarse coat. It writhed, bleated in fear and proceeded to pee all down my best (but heartily disliked) frilly blue gingham dress. I dropped the disappointing creature in disgust and ran for home!

One would think that the lamb episode would have taught me a lesson; but no, on I went undaunted until a suitable time arrived for me to proceed with my second and most important ambition which was to be able to whistle loudly through my fingers just like the shepherds at the Norton sheepdog trial.

By then, we were living in 'Little England Beyond Wales' (as it was known) or Pembroke Dock, South Wales, as it is defined on the map! School was a 3-mile walk over 'Barrack Hill' which gave me an aerial view of the peaceful setting of Milford Haven. Often I would pause to watch the breathtakingly beautiful sight of a White Sunderland Flying Boat skimming across the calm water until finally and effortlessly it ascended against the picturesque backcloth of the blue Prescelly Mountains. It was from these same mountains that the famous blue stone pillars of Stonehenge in Wiltshire were hewn.

It was high up on Barrack Hill where I practised my whistling, all alone except for the somewhat surprised sheep. It went on and on for weeks, me blowing through my fingers until I was dizzy and, on arrival at school, being sent in disgrace to the headmaster's office for continually arriving late. One damp and misty morning, it finally happened. I actually managed the tiniest of whistles—I was so surprised I could barely believe my ears. Within days, my attempts grew louder and shriller. I could imitate the birds, whistle tunes, use two fingers or four and even manage a 'baritone' through a space between my thumbs. At long last I could whistle!

Twenty-seven years on, despite the many jibes about a 'whistling maid and a crowing hen', I am still whistling. Apart from being necessary, in order to work a dog over a great distance, nobody can deny that it is also an extremely useful way to attract the attention of a certain shepherd and small boy when 'dinner is served' (my husband and son). If they fail to appear, a dog or two will always arrive instead!

All through childhood one of my most treasured possessions was a copy of *Black Bob*, the *Dandy* comic Wonder Dog. This book is about the exciting adventures of a working Border collie. Although it is frayed at the edges, I still have it today, tucked safely away in a drawer and only brought out as a special treat when little boys are good.

Nowadays, during the summer months, we often pass through the lovely Border town of Selkirk, our car overflowing with collies (and looking more like Noah's ark), en route for the many sheepdog trials in the area. I am reminded that, for me as a child, this was where the action was. Black Bob and his master, the shepherd, Andrew Glenn, put Selkirk well and truly on the map for me.

Chapter 3

Employment I

A friend of my aunt was responsible for finding me my first job on a farm. The fact that it was, to quote, 'in a wild and lonely spot', did little to dampen my enthusiasm. Letters of introduction were exchanged and subsequently it was arranged that I would be given one month's trial.

I—and a few practical belongings—was to be collected at a nearby town. Determined to be punctual and filled with excitement, I hurried to the rendezvous with my mother. 'He' was already waiting—a dapper, wiry man wearing spectacles. His hair was auburn and crinkly and his complexion was ruddy and lined through squinting at the sun. Being a Yorkshire man, most of his sentences were punctuated with 'nay' and 'naw-but' and his speech was slow and his words well thought out.

The introductions over, I said my 'goodbyes', climbed into his little maroon and cream van and, with a wave to my mother, sped off into the unknown. After a speedy journey through breathtaking countryside, we finally drove up a steep and bumpy lane which caused the little van to vibrate so much I thought it surely must shake to pieces. Suddenly we were met by an awe-inspiring sight for here, built of mellow stone, set snugly into the hillside, was an old farmhouse over-looking a green and peaceful valley. Towering in the background were the Yorkshire moors, dark, mysterious and inviting.

The family—his wife and two small sons—came out to greet us. One boy was

shy, the other grinned from ear to ear. They were as different in looks as chalk and cheese—Colin was quiet, sturdy and fair, a gentle boy with his mother's eyes whilst Lennie, chubby and red-cheeked, had round, bright 'forget-me-not' eyes.

Within the hour and after a scrumptious meal, I had settled in and unpacked my few belongings. At bedtime, the boys showed me to a room next to theirs which had a lovely view of the moors and, within moments of sinking into the billowing feather mattress, I was sound asleep. The following morning, I awoke to the sound of clanking pails and crowing cockerels. My life on the farm had begun.

There were several collies there and I soon befriended Moss, an elderly dog with more character in his face than any dog I've met since. Moss honoured me by his limping presence as I explored the surrounding moors for, until our meeting, he had been a 'one-man' dog. Perhaps he thought I needed a chaperone.

As the glorious April days of 1959 passed by, one by one, the local batchelors put in an appearance at the farm. It was Dick, the local roadman-cum-shepherd and his ebony bitch, Auld Lass, who became my special friends. Although more than twice my age, Dick showed me great kindness and respect, patiently teaching me about country life and the ways of sheep.

At the end of the month my employer asked if I would stay on and, in lieu of wages, presented me with a pair of stout leather boys boots to wear on the hill. Although my badly blistered feet bled in protest at their 'starched' presence, I refused to take them off, firmly believing that without them I could not call myself a true shepherd in every sense of the word.

One day it was suggested that I should have my own sheep dog and so, after a trip to 'pick a pup', Kim came into my life. She was bare-skinned, black, white and tan and I doted on her. Little did I realise at the time that our relationship would be short-lived.

One balmy summer's eve, high on the sunny open moor, Auld Lass pointed out a sick ewe. A Swaledale of good breeding, her udder was blue and painfully swollen with mastitas. I elected to stay beside her whilst Dick, with Lass at his heels, hurried off home to fetch the tractor and transport box. When the sun went down, it became cold and dark and I snuggled up to the ewe for warmth.

After what seemed an age, I suddenly became aware of a bright orange glow in the night sky. Curious, I scrambled through rough heather on to high ground to gain a clearer view. I will never forget the sight that met me for, stretching as far as the eye could see, the moor was a burning inferno. Acrid smoke choked my nostrils as the flames leapt skyward, dancing and flickering higher and higher. We had had a particularly hot dry summer that year. All country people dread moor and forest fires which can occur so easily through the carelessness of the unwary, and on that August evening, the conflagration before me was like the end of the world. For a while I stood there mesmerised, bathed in firelight, until I was brought to my senses by the roar of the little tractor engine, throttle open wide, wheels churning up the ground as it raced to our rescue...and not before time.

Two weeks after this, a severe and prolonged attack of jaundice ended my memorable stay at the cosy little farmhouse in the lovely undulating Yorkshire moorlands and I went home to recuperate and plan my next adventure.

Chapter 4

Employment II

After a period of convalescence lasting several weeks, I decided to see life further afield and placed an advertisement in a popular farming magazine. Within a fortnight, protected by a thick layer of puppy fat and brimming with enthusiasm, I sallied forth once again, this time by train to west Wales sadly minus my devoted bitch, Kim, whom I was requested to leave at home. To quote: 'We already keep two working dogs—quite enough for a small farm.' My good friend Dick therefore offered Kim a home, promising to send her on should my new employer change his mind.

After a long, slow journey south and then west, I arrived at the railway station in darkness. It was pouring with rain and the platform was cold and empty. Fortunately, the buffet was still open so I went inside to warm myself and drink a steaming cup of tea. I emerged ten minutes later to find an elderly, stooped gentleman with long white hair anxiously looking about him. On seeing me, suitcase in hand, he came over and introduced himself in an extremely cultured

I have come to admire felines, as my doorstep bears witness!

voice and enquired about my journey as we proceeded to the station car park where he had left his wife. I was greeted in a warm and friendly way by this gracious lady and I liked her immediately.

After securing my suitcase, we all climbed into the car and drove off into the wet night. The lady and I huddled closely together, our knees covered by a thick blanket. Out of the darkness, my future employer enquired: 'I do hope you like cats.' Having been brainwashed by Mother not to, after two stray kittens soiled her bedspread, I mumbled: 'Er, not really.' However, since that evening I must admit that I have come to admire most felines, our door step bears the proof!

We started a long climb in low gear up a bumpy lane. (Was my life to be filled with bumpy lanes, I asked myself?) Finally we slithered to a halt. 'Well, here we are, dear, home at last!' exclaimed the lady of the house. I leapt out, landing on soft, muddy ground and was more than a little surprised at the incredible sight before me—a bright oil lamp had been placed in welcome beside what could only be described as a broken and battered back door, completely devoid of paint and, milling around our ankles, mewing, pressing, pushing and purring, were a delighted throng of cats. What on earth have I come to? I wondered as they tickled my bare legs with their wet fur. 'There must have been a million of every shape and size!' The words of the song flashed through my mind, and I giggled somewhat hysterically.

I was taken inside and given a hot drink. Afterwards a candle was placed in my hand and I was shown to my room. It was dark, damp and dismal. I slept fitfully. The following morning at breakfast I was introduced to the only other member of staff, Elizabeth, who I was to replace just as soon as she had shown me the ropes.

Translated from Welsh into English, the name of the farm was 'Hill of birds'. To me, even to this day, it will remain 'Hill of pussies' or 'Pussy hill!' for they really were everywhere one looked. In the milk jug, in the sink, in the custard and often, much to my horror, squatting in the corners. On these occasions, I would say in a loud voice: 'What is pussy doing?' but it was all to no avail and so I would return to chasing cat hairs with a spoon as they whizzed around in my tea cup, pretending to be oblivious to what was taking place around me.

Two of the 'dear creatures' particularly attracted my attention. One was called Mouser because of his sheer ugliness. He was yellow and closely resembled a mangey ferret, both in looks and scent. He was also battle-scarred with non-existent ears and wore a permanent snarl on the left side of his face. At meal times he would glare balefully in my direction through amber slits, every now and then pointing at my plate with a 'club' foot. The other was fluffy Tortie, an extremely beautiful specimen. As her name implies, she was a tortoiseshell and had a flowing, silky coat and luminous green eyes. Tortie bred a mischevious and tailless ginger kitten whom I christened China as he looked very ornamental perched on the window ledge.

One day at lunch I tentatively asked the master of the house if China could be my very own pet. Without a word, he rose from the table and left the room. Once outside, he plucked my favourite from off the window ledge and disappeared with him in the direction of the cow byre. I was never to see China again.

Elizabeth, who shared my life at the 'Hill of pussies' was quite a character with a great sense of fun. We made our own amusement—we had to. Our main source of excitement was provided on washday when, doubled up with mirth, we would peg our frilliest knickers on the line, either side of our strait-laced master's long, woolly underpants.

At meal times we were regularly reminded by him that he disliked greedy girls, with the result that we ate little. Every Friday both he and his wife had an 'away-day'. The moment they departed, we sprang into action and baked furiously and then hid the goodies under beds, in suitcases and anywhere else that came to mind. During the rest of the week, if we felt peckish, we would creep quietly into each other's rooms to partake of midnight feasts. I'm positive it must have been the crumbs which brought about our downfall for, in no time at all, I was deemed capable and able to cope on my own and thus my friend went off to pastures new.

For a long while after she had gone I felt extremely lonely. At Christmas I was at my lowest ebb. In order to cheer myself up I decided that, after milking on Christmas Eve, I would open the gifts I had been saving. So, on the stroke of 10 o'clock, by the light of a flickering candle, there I sat in all my glory, on a damp and steaming bed surrounded by a mountain of gift paper, sipping sherry from a cracked cup without a handle and stuffing myself with dates and marzipan. From that day onward I stopped feeling sorry for myself and determined that I would try to see the funnier side of life, whenever possible.

The farmhouse was large, gloomy and extremely eerie and, with only a candle which constantly threatened to blow out, bath nights were a real nightmare. I can still feel the hot wax dripping on to unsuspecting fingers as I made my way up the dark and draughty stairway to the boxroom-cum-bathroom. The cause of my fear was a huge bat, a terrifying creature which often frequented the place. He would hide high up in the rafters, lurking in the shadows until I stepped, naked and trembling, into the water. Then, just like Count Dracula, he would swoop swiftly and silently in my direction. The only thing I could do was submerge quickly and only surface for air when I felt my lungs bursting. The first time he put in an appearance, I let out a shriek loud enough to wake the dead (I thought). It didn't even wake the living!

As time passed by I got used to George, as I called him, and his peculiar anti-social antics. I realised that he was just inquisitive, and discovered that if I blew out the candle he would leave me alone. Perhaps he was cold . . .

It was at the turn of the year that the accident happened. Each evening, before milking, I would sit at the kitchen table priming the two oil lamps, one for each end of the byre. On this particular evening the mantle of the first lamp began to glow and so I removed the butterfly clip which, when soaked in methylated spirits and ignited, was used to light the lamp. I blew on it as usual but failed to notice that it was still alight as I popped it back into the jar. There was a loud bang and a flash as it burst into flames. I leapt backwards stepping on poor Mouser who let out a terrified meow and raced for safety. Suddenly the green serge tablecloth burst into flames. Quickly, and much to my admiration, the lady of the house took hold of it by the corner and trailed it out of the kitchen, down the tiled passage and out of the back door.

Panicking cats fled in all directions and not one, after the fright they received, dared venture back over the doorstep, much preferring the safety of the outside window ledge where they sat wearing disgruntled expressions and badly singed whiskers.

Within the space of a week, but not due to shock, the old gentleman fell ill and took to his bed. The doctor was called and Brucellosis was diagnosed. I was thrown in at the deep end overnight. The whole responsibility of the farm, its cows, calves and Llanwenog (Welsh Blackface) sheep became my responsibility and remained so for the following long and arduous six months.

I had to work as I had never done before—up each morning at 5 o'clock—just in case the Lister engines failed and I was forced to milk by hand—then wash the machines before taking the milk by tractor to the end of the lane where it was collected by a lorry. All of this was completed before breakfast at 9 o'clock. Oh, was I ravenous by then! On top of these tasks, there were calves to feed, sheep to look after and hay to cut, truss and bring home for the cows. At night, there was the same milking procedure as the morning and then, after mucking out the byre, my final chore was to refill the hods with coke, a job I dreaded. With only the oil lamps to help me I had to define which was coke and which was pussy poo . . .

To crown everything, the Hydram responsible for our water supply decided to go on the blink. It took two weeks for the spare part to arrive. Two weeks of wet back-breaking work as I toiled time and time again up a steep and slippery incline, carrying two three-gallon pails of clean water. Weeks turned into months. The poor old gentleman, gravely ill, tossed, turned and sweated indoors whilst I sweated out of doors.

Thankfully every cloud has a silver lining and it was during this time that I struck up a marvellous friendship with the two farm collies—Bob, a small hairy fellow with large pleading eyes, and Ben, his bouncing black offspring.

Bob worked for me from the word go, except when it poured with rain. On such occasions he would walk out on tiptoe then beat a hasty retreat back to his bed. Ben, though yet untrained, was full of enthusiasm. Secretly I relished the thought of training him myself. Dare I? Yes, of course I dare.

At the end of three months Ben was extremely competent and was working like a dream, a natural born sheepdog. Gradually the old gentleman gained strength. The day I was dreading finally arrived. I was told very politely that it would no longer be necessary for me to exercise his dogs. The hardest part was yet to come. Bob and Ben, by now used to accompanying me everywhere, flatly refused to follow their master. I was told to discourage them by sending them away. I had no other choice and as long as I live will never forget the hurt, surprised expressions on their faces.

Perhaps, lurking in the back of my mind, was the faint hope that I might be allowed a dog of my own and on one occasion, I tentatively broached the subject, but it was to no avail. Finally, after a great deal of thought and with the hope that somewhere there was a kindly farmer who would allow me to keep a dog of my own, and much to the old gentleman's surprise, I handed in my notice.

Chapter 5

Employment III: With Ernest Crisp

During our life most of us come across a character who stands out from the rest. Because we are impressed by him, we tend to mimic his ways and sayings. Very often, unintentionally, we adopt certain mannerisms which remain with us until the end of our days. Ernest Crisp represented such a character to me. He was my third employer and the first to allow me to own, not one dog, but many. I originally went to his farm to assist with the lambing but ended up staying for 12 enlightening years, employed as his shepherdess-cum-'Jill of all trades'.

It was Ernie who gave me Kim II, a rough-coated, black and tan version of the former Kim who I had left with my friend, Dick. 'Mind ye'll nivver get her to stop', he warned, and I never did either, except when *she* was dog-tired!

One morning Kim ran off and mated with her dad who lived on the next farm. His name was Rob and he spent his spare moments biting quarry men, first having knocked them off their bicycles. His busy times were spent mostly submerged in the water trough which reminded me of the times when I hid to avoid George, the bat. Rob was merely avoiding what he'd been bred for—work.

A bonny dog pup came of Rob and Kim's union. He also bit people but the compensation was that he was a great worker. I called him Shep and he became my first trial dog. On his first outing he somehow managed to pen the sheep in the Secretary's tent which didn't go down too well with the other occupants! Unfortunately, poor Shep, due to his close breeding, had an unusual affliction for a dog. He was cross-eyed and I often wondered just how many sheep he actually saw.

In those days, I often travelled to trials by bus with my crook and Shep safely hidden under the seat in front. He didn't care much for this mode of transport and it was no wonder that he was often in a dreadful state of nerves by the time we reached our destination. Shep and I never won a prize, but we certainly had a great time.

In all of those years I knew Ernie, I witnessed him lose his temper on only one occasion and that was when my collie, Shep, bit him on the ankle! It would not have mattered, according to Ernie, had it not been for the fact that the dog chose to bite him on his own doorstep!

Although neither he nor his wife were sheepdog trial enthusiasts, they regularly accompanied me to trials, giving me every encouragement possible. As the years went by I grew to like and respect Ernie, mainly for his honesty and integrity, but most for his great humour. I found his oft-repeated jokes and comical quotes hilariously funny. To him a crane fly or 'daddy long legs' was known as a 'spinning jenny', a plover was a 'pee-sweep', a wood pigeon was a 'cushit' and the plaintive curlew was aptly named a 'wharp'.

As Ernie was one of the old school, work had to come first. He was a firm believer that you only got out of both life and land what you were prepared to put into it. He stood 6 feet 1 inch in his size 12 tacketty shepherds' boots, with their turned up toes—'all the better for rocking uphill', he would say with a smile as he adjusted his weather-worn cap to an even jauntier angle. He was a lean and wiry man, with a slight stoop, apple-red cheeks and deep set blue eyes who ran his 300 acres in the old-fashioned manner in which he himself had been raised.

When I first appeared on the scene as a very inexperienced 18-year-old, one of my first jobs that autumn was to learn how to cut corn and service a rattling and ancient binder. I can still recall my feelings of horror as I proceeded gingerly up the field with wooden lathes pinging off the binder canvas because no one had thought to instruct me in the art of tightening it up!

In the winter months I endured long and wearisome hours, forking sheaves of corn up to the 'feeder' on top of the threshing machine. How I detested the dust, noise and the scurrying rats which were ever theatening to escape up the insides of my trouser legs, until I finally realised why everyone else tied binder twine

Right *Kim II and Viv Billingham.*

Ernest Crisp.

below the knee! Regardless of what strain of oats we drilled in the spring of the year, like Topsy they 'just growed and growed'.

In the late summer, a wild Galloway bullock leapt over the fence into 40 acres of 6-foot high oats and went missing for two whole days, finally emerging extremely thirsty but otherwise none the worse for his ordeal.

'Fattening beasts' were well into their third year before the word 'market' was ever even mentioned. Once there, if the price was not right, back home they returned for a further six months of feeding. The hoggs, or yearling sheep, were folded together and fed on measured patches of turnips—which we called 'turnip breaks'. In the late afternoon they were transferred to the next field where they spent the night on clean grass before being fed corn in the early mornings. On such a diet the hoggs became ready for market very speedily!

I spent many a day 'shifting the sheep nets', often in cold frosty weather, so that my fingers became numb and fumbling and, when 4 o'clock came, I was glad of the chance to rush away and 'corn' the ever-hungry ewes. Then, with darkness rapidly descending, and the twinkling stars beginning to appear, clear cut, suspended in the deepening winter sky, I would run home, dogs dancing around my heels, to do the evening milking before feeding them and settling in for the night beside a glowing log fire.

In fine spring weather, it came as no surprise to find Ernie sowing grass seeds all through the night. He used an old fashioned 'fiddle' hung around his neck on a canvas strap and would stride up and down with left leg and right arm moving rhythmically like a ghostly musician, with only the moon to guide him on his march to the headland! I can see him quite clearly, resting on one knee while his ever-patient wife filled the empty seed bag for him. At 7.30 the following morning, she would be relieved of field duty to milk her two house cows, June and Mary-Anne and to give the calves and poultry their breakfasts before she attended to her many household chores.

In Ernie's later years, his eyesight began to deteriorate. After a great deal of persuasion, he paid a reluctant visit to an eye specialist who informed him that, if he went on smoking a pipe, his sight would 'worsen'. Even after this ultimatum, he continued with the habit (many's the chiding I gave him because of it). He informed me that it was very difficult to give up what had become his main pleasure in life so he would smoke when no one could see but, owing to the strong aroma of tobacco on his clothes, his very caring wife began to have an inkling about these 'secret smokes'. She was concerned because he assured her that he had given his pipe up, and so she asked me if I would keep an eye on him and report back.

I would discover him, quite unintentionally, in the most unlikely places. Rather like a mischevious schoolboy, his eyes would twinkle as he endeavoured to hide the offending pipe and waft the smoke away with his hand. Even though I worried dreadfully about his eyesight, I could never bring myself to tell tales. Having been regularly caught out by his wife, he smoked less with the result that his eyesight improved.

Sadly in 1978, at harvest time, Ernie passed away, after a brief illness. He had retired recently to a stone-built cottage on the outskirts of the village where he was born and, after all his years of hard work, I considered that he deserved much more than his allotted three score years and ten, but it was not to be and so I lost my dearest friend.

I visited him in hospital shortly before he died and sat at his bedside. He could see that I was saddened by his appearance and he took my hand in his large work-worn ones, holding it tightly, comforting me. When it was time for me to leave, no words were needed other than a whispered 'goodbye' for we both realised that this would be the last time we would see each other. I paused in the doorway for a final look at the man who had moulded me in his ways. He had a far away look in his eyes, no doubt dreaming of times gone by.

Once outside the hospital and warmed by the autumn sunshine, I couldn't help but smile through my tears as the words of one of his favourite ditties flitted through my mind:

The Farmer stood at the pearly gates,
his face was scarred and old.
He'd come to gain permission there
for entry to the fold.
'What have you done,' Saint Peter asked,
'to gain admission here?'
'I've been a farmer, Sir,' he said
'for mony and mony a year.'
The pearly gates swung open wide
as Saint Peter rang the bell.
'Welcome, and come in,' said he,
'You've had your taste of hell!'

Ernie had been born at the end of October, on the eve of All Hallows (or Hallowe'en). His father had caught rabbits for a living. He had married Ernie's mother when she was over 30. She was 'in service' at that time, employed as a cook and working in a 'big hoose'. She bore him six children in quick succession. The first born, Robert, was killed in action during the First World War. Then came Samuel, William, Ernest and Mary, in that order. Sadly, another daughter died in infancy after swallowing a half-penny piece.

Ernie's first childhood memory was a painful one—being inquisitive he inadvertently shoved his forefinger between the rollers of a mangle as his mother turned the handle and, although he quickly recovered, his finger was never quite the same again—from then on it had a crooked squashed appearance.

Ernie heartily disliked school, or 'skee-ool' as he pronounced it. This was probably due to the fact that the master heartily disliked Ernie, who had nicknamed him 'The German Spy' because of his accent rather than his origin! One Friday afternoon things finally came to a head and Ernie, a big lad for his age, thumped the angry master on the end of his nose, then, deciding that he'd had enough schooling, he went hill-lambing instead.

Ernie would relate these stories to me over and over again as we sat cross-legged on the dusty floor of the upstairs granary, mending corn sacks on wet and stormy afternoons. Occasionally he would pause and push back his 'pee-sweeps tail', a long piece of hair cultivated especially to cover up the bald patch on the top of his head. Following that he would grin impishly showing his few remaining tobacco-stained teeth before continuing on with some other yarn, or song, his favourite being *I'm gonna marry a shepherd laddie:* 'Aww, I wouldna marry a Shepherd Laddie, far too mony doggies. There's only Tip and Tam and Toosie, Dick and Nan and Susie, you wouldna call them mony doggies?'

Sadie, Ernie's wife, was a farmer's daughter. He'd courted her for quite a while, on his motor bike! In those days, she wore long black stockings, short skirts and a shapeless hat. One night he asked her down the garden to take a look at his ferret. She liked him (the ferret) so Ernie proposed. They lived happily, though were childless but, as Ernie pointed out: 'We've nane tae please us and nane to vex us either.' His main regret was that he hadn't bought the farm, and the next one to it! 'Land was dirt cheap after the war,' he would say.

Ernie's complexion was flawless. Each evening he would remove his collarless striped work shirt and stand with braces dangling at the kitchen sink. Then he would scrub himself all over with household soap. Although a stickler for clean-liness, he didn't care much for his appearance, except on Sunday when he smartened himself up to perform his duties as sidesman at St Mary's church. During the week he walked about in any old clothes. Many's the time when visiting representatives would enquire where they could find *his* employer. Often he would say 'over there', giving me a wink and pointing in my direction with his crooked finger.

At lambing time, he presented me with a half bottle of whisky—'for the weak lambs, warm it in yer mouth and spit it intae theirs'. So I tried and tried and finally there was very little left and I felt very queer, but happy. It was a bitterly cold day and I was warm and glowing both inside and out. Unfortunately, at that inopportune moment, the Wages Officer chose to arrive with the intention of asking me if I was satisfied with my lot. Under the circumstances, I did the only thing I could and shot off on all fours into the maze of passages that connected the farm buildings. The poor man, believing that he'd frightened a young and innocent girl, finally cornered me in the cattle crush. Luckily by that time I was beginning to sober up and so answered his questions coherently. What he thought I'll never know. After his departure Ernie took one whiff of my breath and stood over me until I had devoured a quarter pound of peppermints—'just in case Sadie finds out'. Needless to say that was the first and last bottle of whisky I received 'for the lambing'.

In addition to being a sidesman at the church, Ernie cut the cemetery grass with a scythe and provided produce for the harvest festival. As this time approached, we would busy ourselves scrubbing potatoes and turnips, but it was the preparation of the corn which I enjoyed the most. I would watch Ernie and think: 'this is how I'll remember him in years to come'—sitting in the corner on a bag of binder twine with gnarled hands putting together neat bundles of corn, strong fingers running through each bundle stripping the straw from the stalks, tying the bundles top and bottom before cutting them to the correct length with a pair of rusty sheep shears.

It was Ernie who taught me to hand-shear sheep. The 'out-bye method' he called it. First you clipped away down the right-hand side, starting at the back of the sheep's neck and carefully keeping her 'lugs' out of the way with your other hand. After clipping the right-hand side, you would step over your fleece at the same time hoisting the often protesting sheep up on to her rump before proceeding down the left-hand side.

Ernie and Sadie Crisp with my son, Geoff Jr.

Shearing sheep.

Ernie could clip with either hand. I could only clip with my right. Consequently, his sheep always had a better look than mine. After a couple of seasons, fool that I was, I began clipping more sheep in a day than Ernie. When he finally noticed, he downed tools with a clatter muttering: 'When the pupil gets better than the teacher, it's time to quit.' And quit he did so I was left to clip the lot from then on. There must be a moral there somewhere.

Most winter weekends found us sawing logs for the fire. We used a cross-saw. Innocently I asked why he didn't use a power saw. With a smile he explained that by using the cross-saw, we got two warmings for the price of one.

One day Ernie's sister, Mary, invited us all to tea. A rare treat for me. She had bought a new carpet and they had just installed central heating in their home. Although the weather was warm, it was turned on full and all the windows were open wide to let out the heat (or let in the cool, we couldn't decide which). Ernie had guessed that we had been invited so that she could show off.

The table was covered with a lovely lace tablecloth and laden with scrumptious goodies for Mary, like her mother, was a good cook. We sat down to eat. Suddenly, a large, fat cat appeared from nowhere and leapt on to Ernie's lap and promptly dug her claws into him. Now Ernie hated cats and his reaction was severe to say the least. 'Get off you brute!' he yelled, attempting to knock the somewhat startled cat to the floor. Filled with alarm she stuck her long, sharp claws into the lace tablecloth with the disastrous result that both she, the cloth and everything on it, landed in a sticky, messy, broken heap on the new carpet. It was a long, long while before we were invited again.

A few weeks later, Sadie and I had a bit of excitement. Clad in our best bib and tucker, we set off for London so that I could take part in the BBC television programme *What's my line?*. (A shepherdess in those days apparently was something of a novelty.)

Eamonn Andrews was in the chair and Barbara Kelly, Lady Isobel Barnett, David Nixon and Marshall Pugh were on the panel. During the rehearsal I accidently left the stage in the wrong direction, colliding with the next contestant. 'You won't do that on the actual programme?' enquired the worried producer. 'Oh, no,' said I. (You've guessed—I did!)

While I was awaiting my turn, a man opposite kept staring at me. Thinking that well brought up little girls should look the other way, I did. Later on I found out it was the guest artist. His name was Harry H. Corbett of *Steptoe & Son* fame. I wish I'd said 'hello'. Anyway, when my turn came I managed to beat the panel. David Nixon asked did I make jam roly poly? (For my mime I rolled up a fleece.) The following day, after a quick trip around Madame Tussauds, my diploma to say I'd beaten the panel safely packed in my suitcase, we set off for home. Needless to say, our 'excursion' became Sadie's favourite topic of conversation for the next ten years.

We were threshing corn when poor Ernie fell through the granary floor. I followed him in and was just about to carry my burden up the staircase when I was surprised to see a long thin leg with a large tacketty boot on the end of it, dangling through the ceiling. He had been carrying a 12 stone bag of barley when it happened so it must have been quite painful. Anyway, we lifted him out and, apart from walking a bit strangely, he didn't show any ill effects. That evening over supper I asked if it had hurt. He replied: 'No', so I said in all innocence: 'Then why were you walking funny?'. There was an embarrassed pause before he replied: 'There's five pounds in the Bank of England for anyone who can mind their own business. . .' (Five pounds was a lot of money in Ernie's day!)

I still miss the old fellow a lot.

Chapter 6

Working at Swindon

Meg, Jed and Trim

Geoff Billingham and I met at a sheepdog trial, and in all honesty I must confess to being attracted to his lovely bitch Meg long before her master caught my eye. I suppose one would call it 'love at second sight' with Geoff and I, for it was some years before we were married. Geoff (or GWB as he signs his sketches) achieved his boyhood ambition in 1973 when we came to live in the Scottish Borderland, 'Sheepdog Country', with our young son, Geoffrey William ('Geoff Junior'), then a babe in arms. It was on a bright October day when we unloaded our possessions and animals on Cheviot soil and made Swindon House our home.

Meg was a competent bitch who had an unmistakable intelligence in her luminous eyes. Bred by Bob Short out of Gay and sired by Tom Watson's Jeff, Meg was incredibly wise and devoted to one master. She came into my husband's hands at the age of five months and grew into a medium sized bitch, prettily

Above *Swindon in winter. Geoff Jr with pony, Dinah, Laddie (left) and Garry* **Below** *Meg at work.*

marked, with semi-erect ears and a soft, dense, watertight coat. During the first eight months of her life Meg showed no interest whatsoever in the sheep which she had been bred to work. One evening, whilst out for exercise, she decided that the time was ripe and slipped away to gather a flock of ewes, bringing them to her master's foot, as though she had done so every day of her life and, from that day on, Meg never looked back in her career as a sheepdog.

As a worker she was beautiful to watch—very smooth with a quiet but authoritative method. She always followed her sheep at the correct speed and distance and could adapt to their ways so as not to upset them. Her outrun was well thought out and careful, drifting in steadily to the point of balance so that she could weigh the sheep up and lift them gently, often without a command from the time when she left her master's side until she returned to him with her charges. The most unruly sheep quickly settled to her quiet approach.

Meg excelled herself at the pen. She would look obstinate sheep straight in the eye, all the while gently, but firmly, backing them in when necessary. During her seven trial seasons she won and was placed on numerous occasions. At the Chester International, she gained a fifth place in the shepherd's class after a smooth, flowing run.

Normally after running at a trial she would lie quietly in the background, occasionally wandering over to the refreshment marquee in the hope that some-one had dropped a sandwich or a bun by mistake. We never chained her up and the one and only time that she gave us cause for concern was at the Chester International. After her run she completely disappeared. We searched everywhere without success until she was discovered fast asleep underneath our car! There were hundreds of cars parked close together. We had only owned this particular car for two days. Meg had travelled in it for the first time to the Chester International trial!

Unfortunately, Meg was not a prolific breeder, only producing three small litters in her lifetime. Unless she approved of her 'would-be' mate, she would chase and attack him in a ferocious manner. The few puppies which Meg produced, although intelligent and nicely marked, often proved to be shy and sensitive. Bett came out of Meg's first litter. She was a smart, almost black bitch. A powerfull triallist, coming in fifth in the shepherd's class at her first National, she was sold to the late Stan Harrison but she died young, of cancer, after being kicked by a bullock.

We sold all of Meg's second litter when they were puppies, because we had too many other young dogs at the time. Her third litter produced Jed, Trim and Glen—a sound, dependable work and trial dog. At ten weeks of age Trim was a natural worker. By the time Glen reached the age of six months, we could do all our shepherding jobs with him.

Jed was my pup. She was extremely timid and shy and very nearly died at birth. I delivered her on the back seat of the car, on a return journey from the vet. I had taken Meg to the vets because owing to her advanced years we expected a prolonged labour, I worked on tiny Jed for almost an hour before her breathing became regular and she was ready to suckle her anxious and ever-watchful mother on the back seat.

Jed, GWB and Trim (Kelso Chronicle, Tweedale Press).

By weaning time Jed would follow me at a safe distance. The moment I reached out my hand to touch her she would back away in fear. When she was three months old, we took her to visit a flock of Swaledale ewes. Her visit resulted in absolute chaos, with the sheep bolting in all directions. She refused to come to either GWB or myself and although we made numerous attempts to capture her, these proved fruitless. Finally she became thoroughly exhausted and we were able to use her mother as a decoy to attract her attention while we got hold of her.

A couple of months went by and Jed still refused to be handled and so, reluctantly, we gave her away as a pet to a friend in the town. After a month we received a telephone call to say that our friend's circumstances had changed; he could no longer keep a dog and would we like to take Jed back. We set off to fetch her and were pleasantly surprised to see how friendly she had grown, living in a

house. We were more than happy to return home with a less inhibited collie fast asleep on the back seat of our car.

As time went by Jed became a replica of her mother but was more willful. Ignorantly we put this down to her youth and inexperience but she did not mellow with age.

Meg grew slower in her old age. She would set off for the hill with her master and the younger dogs, quickly realise that the journey was beyond her and slowly and forlornly return home to me. Independent and cheerful to the end, her death came suddenly and unexpectedly. We found her one winter's morning in the byre, lying peacefully as though asleep.

Her daughters Jed and Trim have gone on to great things. She would be proud of them for, between them, they have won and been placed at trials all over the country. They won the brace class at the Scottish National and represented Scotland at International level on three occasions. They came second at the Bala International in 1980 and in 1981, they became 'Champion of Champions', running brace at the Royal Welsh Show.

Like their mother, Meg, neither bitches have been prolific breeders. However, the puppies that they have bred have made sound work and trial dogs. Hemp and Ben, two powerful sons of Jed, muster large flocks in New Zealand while a daughter of Trim—Whitelaw Ceilidh—works and plays in sunny California.

Garry
Garry came into the world on a bright May morning within sight of the mist-

Left *Father and son with Trim and Jed* (Frank Moyes).

Right *Jed, the Sourhope Dog.*

shrouded Cheviot hills, two weeks after his mother, Jed, had finished her busy lambing stint in the steep reclaimed fields at Swindon. He was the smallest of a litter of five puppies, sired by another Jed known locally as 'The Sourhope Dog' (after the farm where he lived, not because of his temperament!). Sourhope Jed is now beginning to show his years, his old legs are getting slower, although his appearance remains unchanged. His coat is glossy and his eyes still shine. He is a sturdily-built, cleanly-marked, curly-coated dog with a white head, whose power, line and quiet method of handling sheep had first impressed GWB as he watched him working a flock of Blackfaces on his home ground, from the steep east side of the Whitelaw hill. Jed's sire, the half white-faced Cap, had been even more impressive in years gone by. His pedigree contained the blood lines of J.M. Wilson's famous war years hill dog—Cap, Kirk's Nell and a great deal of Bob Fraser's powerful 'Mindrum' blood.

Garry quickly grew into a medium-sized dog with a luxurious thick curly coat which protects him during the cold winter months. He has a neat cleanly-cut head with a wide brow and a good stop. In profile, his skull is noticeably long and flat along the top. His markings are clean—jet black and white—with a full white face blaze extending through between his ears to join a wide white collar. Garry's nicest features are his eyes. Large, luminous and full of expression, bright orange-brown colour, they blaze like twin fires out of his honest face.

From a very early age there was a noticeable quality about Garry. It was obvious in his stance, the way he held his head and his calm attitude to life, inherited from his sire, and his maternal grandmother, Meg. Unlike other young dogs we had owned, he never attempted to cross his course when sent for sheep. He would gather sheep from any distance and he never threw his tail over his back when it was necessary to put in a tight turn.

I find that Garry is a most devoted and intelligent worker and companion, in spite of me and not because of me! I chose him on the day that he was born because I liked his markings for he bore a strong resemblance to Black Bob, my

Above *Garry—a real old-fashioned dog.*

Left *Jed's litter. (Left to right) Duke, Tweed, Ben, Garry and Lark.*

Below *Garry bore a strong resemblance to Black Bob, my childhood canine hero.*

childhood canine hero. There were four well marked dog pups in the litter, and a bitch with a lot of white on her whom we christened 'Lark'. At the time of writing she lives further on up the valley where she helps her master to shepherd the bleak windswept Cheviots.

Besides Lark and Garry, Ben, Duke and Tweed (GWB's pup) made up the rest of the litter. Duke is trialling down in Yorkshire. Sadly, Ben's new owner had him put to sleep after a broken limb refused to heal and Tweed died of Leptospirosis of the kidneys when he was two years old.

When Garry was a young puppy, he rarely joined in the fun and games with his sister and brothers, much preferring to sit and solemnly survey what was happening around him. He was a real old-fashioned type of dog and, for such a tender age, he was exceptionally faithful, refusing to follow anyone but me, and this he did so closely that he was constantly underfoot. As he grew older and gained more confidence, he gradually increased the distance between us. If he managed to get too far ahead, he would sit and wait for me to catch him up, occasionally glancing back over his shoulder with a nonchalant air.

Most families of puppies are boisterous and regularly get into mischief. There was something special about these puppies, they appeared to be serious-minded and studious, more like older dogs and they did everything together. When they were ten weeks old, I took them across a shallow part of the burn to visit a small quiet flock of ewes in what we call the 'school field'. (So named because of the tiny school nearby where years ago the valley children received their education.) The moment the puppies sighted sheep, down went their heads and tails! They quivered in anticipation and then set off to drive—all in a line like soldiers on parade. The sight of these tiny, fluffy creatures, fully in command of a flock of full grown ewes, held me absolutely spellbound. After a few minutes, I reluctantly called them to my side, afraid for their safety, and we set off for home and supper.

The first hint I got that Garry was of above average intelligence was one morning when I opened his kennel door and he quietly slipped away while I was feeding the poultry. I was not unduly worried at his absence and set off across the fields and over the burn to where my mare, Flicka, was stabled. When I reached my destination, I was pleasantly surprised to find Garry seated on a trailer patiently awaiting my arrival. I had lifted him up on to the trailer the previous morning so he would be out of harm's way. Garry was then aged five months, and to have anticipated that I would do the same the following day, I think was no mean feat for a pup. It showed intelligence and a wish to please.

Due to a shortage of kennels and the fact that he neither stole food from the table nor cocked his leg indoors, Garry eventually came to live in the kitchen. One lunch time, as I was just about to lift my fork to my mouth, I thought I felt a gentle tug upon my sleeve. I continued eating and the tug that followed was so forceful that the food on my fork fell to the ground and was promptly gobbled up by a pup whose eyes were nearly as large as his belly. Since that day, whenever Garry requires attention, or if he feels neglected, he pulls on my sleeve with varying degrees of strength until I enquire what he wants of me, a trait which he has passed on to some of his offspring.

Fortunately Garry has inherited his parents' determination but can, on occasion, when sheep are wild, prove over enthusiastic, pushing them at too great a speed. When he is working close at hand, a harsh whisper is all that is required to get him to 'take time'. If he is working at a distance 'bombing' sheep, I have to call his name gruffly. I find this rather embarrassing, especially at a sheepdog trial, and would much prefer to whistle, were it effective!

When I give Garry a command which he considers is wrong, he hesitates, so I know not to repeat my mistake. On the hill he is quite capable of working with few or no commands at all, though I find that it takes a while to put back the initiative into Garry that running him in trials throughout the summer removes. Running a dog on a small trial course, especially when the sheep are wild, involves a great deal of flanking and over-commanding. In Garry's case, I always feel that it is rather like instructing your grandmother to suck eggs! It is always a relief to get home and out on to the hill after running a dog on this type of course.

On rough ground, Garry is obviously in his element, with so much space around him. A great feeling of peace descends upon me as I watch while he gathers, drives and cross drives until every scattered flock is joined and quietly brought to my foot. However, I have found that too much hill work can cause a dog to get into the habit of stopping well short of the correct 'lifting' position at a trial, as an intelligent dog, when gathering a hill on his own, will push the first 'cut', or group, of sheep he encounters inwards across the hill face before continuing on upwards for the rest of the flock. When gathering a hill I know that Garry relies a lot on scent as he always seems to arrive at the 'right bit' when he goes for sheep, even when they are hidden from view in dense bracken or behind a hilly knoll.

By the age of eight months, he would 'bomb' sheep to me after gathering them from any distance. He had exceptional eyesight and could spot sheep long before I did. As he was extraordinarily mature, or 'old-fashioned' for his tender age, I decided to begin his training earlier than was customary.

I had never had any bother getting him to come away from sheep when asked, and he automatically stopped dead on a whistle. In the fields his outrun was

I feel a gentle tug on my sleeve.

inclined occasionally to be tight, but he swung off the sheep nicely close at hand. He much preferred to work close to sheep and could do so without upsetting them. If there was an occasion when things did not go quite to plan, he always managed to keep his cool and react sensibly until things calmed down.

Opening him out wider on his outrun proved to be a relatively simple task. It was only necessary to stop and redirect him on one occasion and it was learnt. The next step was to teach him the meaning of the command 'take time' and to work on his feet at the pen and shed. Working regularly with fast, wild Cheviot hoggs on the steep hillsides at Swindon taught him to give the sheep more room and flank freely. Because Garry was inclined to be 'strong of eye' it was sometimes difficult to get him to come quickly enough when you required him to 'shed'. After giving the matter much thought (I am a slow thinker!) I tried a 'fast' flank whistle—much to my relief it worked and he has become a useful shedder, although sometimes you do have to ask him to hurry up!

In the early days of his training, he developed the annoying habit of stopping short on his outrun long before he reached the '12 o'clock' position, directly behind the sheep. I think this habit developed through me using him too much on the hill and sending him too great a distance for too few sheep in the fields during late winter. There weren't many sheep because the hill ewes were heavy in lamb and most of the hoggs were running with them, making the gathering of spread sheep an impossibility.

However, I believe the main reason that Garry stopped short was either because of 'eye' or he was afraid that the handful of hoggs he had been sent to retrieve in the fields, were going to bolt in my direction at great speed before he could get behind them, therefore he stopped so as to be in a position to prevent this from happening. Not wishing to confuse or offend him, I spent the rest of the winter and early spring giving him short gathers, whistling him to the 12 o'clock position, allowing sheep to bolt in the opposite direction before sending him to kep (go round) them to bring them out of a corner.

I hoped that these exercises would cure him of stopping short, but they did not. Finally, exasperated after many months, I took him to gather some positioned sheep on a steep hillside where I knew he would stop prematurely. When it happened, I roared: 'What the . . . do you think you're doing?' and was flabbergasted when he got up and raced for the 12 o'clock position. Touch wood, he has never stopped short without a good reason since that day. If I had scolded him when the fault first appeared he would have become confused, because he would not have understood the reason for the scolding. I realise now that prevention is better than the cure (and speedier) so I try not to allow major faults to develop in other young dogs.

Convincing cheeky 'heavy' sheep to walk into a pen has never been a problem for Garry. It is the wild flighty sheep which give him the most bother. He simply takes over as though I was not there and scares the living daylights out of them! My fault again for not doing 'our homework' correctly. I hope I am wiser as far as the next generation of Garrys is concerned.

His ability to 'line up sheep' is above average and, although this is a great asset when driving sheep, it does cause problems with wild sheep at the pen, and when trying to settle sheep in order to shed one off. When Garry was around eight months old, in January, and unknown to me, he slipped quietly away up the steep Whitelaw, picking up a large flock of Cheviot ewes and hoggs on his travels. The sheep had been lying in the shelter of a deep hollow up on the tops, approximately a couple of miles from home. Garry picked them up and brought them down to the house in a line as straight as any arrow, all on his own. (As the ground had a light covering of snow upon it I was able to see the sheep's tracks clearly.) Once he gets on the heavy side of sheep, few or no commands are necessary until you wish to turn them in another direction. Robin Armstrong, owner of 'Sourhope Jed' tells me that Garry's paternal grandsire, Cap, could line sheep up by leaning slightly to one side and pointing with a forepaw!

When I first began training Garry, I made up my mind always to be reasonable in my methods and, whenever possible, to put myself in his place and look at things through his eyes. I was aware that his training would take longer this way but hoped that the end result would be a natural dog rather than a mechanical one. Previously I had not put any real thought into training other dogs and as a result, tended to blame the poor dogs for any faults they picked up rather than blaming myself.

At the first trials Garry and I entered, I had a feeling of guilt as I flanked him all over the place, purely for the purpose of bringing sheep through gates that did not have a fence or a wall on each side of them (especially when he was quite capable of picking up sheep and bringing them to me on his own). I sensed that asking him to flank on the fetch thoroughly confused him. When sheep were picked up at the post, the problem did not arise. Unfortunately, very few sheep will remain where you put them for more than a couple of minutes.

Sadly, 'line' is rapidly being bred out of our dogs purely for the satisfaction of man's urge to flank them about and dominate them in order to win sheepdog trials. As time passed, and Garry grew older, working regularly on wild Cheviot hill hoggs allowed him to flank for a reason, and I sensed his reluctance to flank on the fetch leave him. The correct habit had been formed without upsetting the dog. I have found that, providing I am reasonable with him, he repays the compliment a thousand fold.

Garry is indispensable when I am training 'the young dogs'. Once that he sees that everything is under control, he retreats to the shelter of the dyke where he lies contentedly with an expression on his face that can only be described as half smug, half amused. When I glance in his direction he flicks his ears back and his tail slowly thumps the ground, while his large eyes glow with anticipation. When it is time to go he will rise, yawn in a bored manner, and stretch luxuriously. (He loves a good stretch.) Sometimes he talks to me by making little grunting noises which I interpret to mean that he has not a very high opinion of me as a trainer and could make a much better job of it himself!

His early trials were terrifying affairs for him, owing to his extreme sensitivity. People, cars, trees, buildings, even shadows, affected his performance and often caused him to do things which were completely out of character. He was so afraid of the men who let out the sheep at the top of the field that he would rush off with his packet of sheep without stopping or slowing until he was past or through the

Above *Garry (left) and son, Laddie.* **Below** *Garry and son, Laddie (left).*

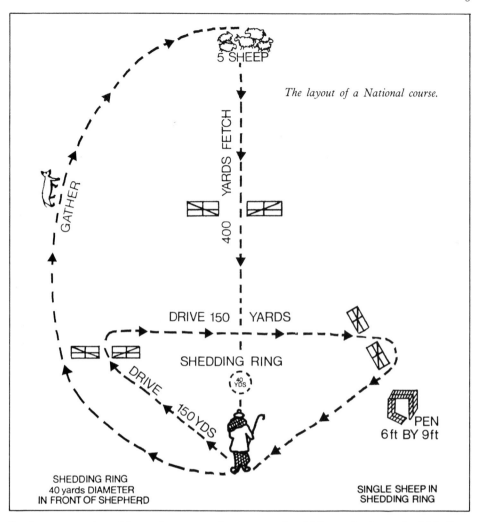

The layout of a National course.

5 SHEEP

YARDS FETCH

400

GATHER

DRIVE 150 | YARDS

SHEDDING RING

DRIVE

150 YDS

40 YDS

PEN
6ft BY 9ft

SHEDDING RING
40 yards DIAMETER
IN FRONT OF SHEPHERD

SINGLE SHEEP IN
SHEDDING RING

fetch gates, and could see the expression on my face! Towards the end of his first season, he became used to everything and began to enjoy himself immensely. These days, he is out at the post and looking for sheep whilst I am still thinking about it!

We competed in our very first Scottish National Sheepdog Trials in August 1978 at Earlston, when Garry was two years and three months old. Apart from a couple of long outruns away from home, I gave him no special training. My main hope on the day was for heavy sheep (and even heavier rainfall to put off the spectators watching our attempt, should things go wrong!)

We were drawn to run on the Saturday. Thursday and Friday dawned blustery and sunny and the sheep were light. I could not sleep a wink on the Friday night, worrying and hoping for a change in the weather. I got up at dawn and gave Garry a long hill gather before breakfast. When we arrived at Earlston, lo and behold my wish was granted. We found heavy, well-behaved ewes (well herded by Tom Liddle, a fellow sheepdog handler) and much appreciated rain, not too heavy but enough to dampen the spectators enthusiasm!

My knees wobbled like twin jellies as I awaited my turn. Garry calmly summed things up, then promptly curled up at my feet and fell into a deep slumber, snoring quite loudly. At last, it was our turn to face the music. The rain descended in torrents. I shakily asked the Course Director, Willie Hislop, whether it was too late to change my mind. (It was!) He kindly walked with me to the post where he promptly abandoned me. We were on our own.

Garry glanced up at my face with calm eyes. I asked him: 'Can you see sheep?' The slight forward movement of his ears told me what I wanted to know. With a quiet, slow whistle, I set him off wide on a right-hand gather. He cast out well from my foot but, on seeing the men at the other end, came in flat in order to get between them and his sheep. The sheep gave a start and bolted to their left, making me realise too late that I should have halted Garry fractionally short on his outrun.

The 'lift' was awry to say the least, the startled sheep running for their lives. Garry only just managed to get them straightened up in order to hit the fetch gates. He then kept them on course until they reached my feet. Then I made another mistake, I should have allowed them to drift off slightly to my right in order to get them turned quietly round about me. Through ignorance, I turned them too sharply, giving poor Garry a piercing stop whistle. The sheep, alarmed, shot off line and were brought back hurriedly. After a reasonable down drive he put them through the gates, managing a tight turn into the cross drive. For me, the cross-drive was the most beautiful part of his run. He tucked himself in on the heavy side of the sheep, head down, needing no commands, until he reached the last gate where the sheep paused momentarily to inspect them before walking through. They turned tightly and were brought in a straight line to the shedding ring.

It was a few minutes before a gap appeared, allowing Garry a chance to come in and shed off the two sheep without red collars and drive them away from the others. This accomplished, I sent him off to head them, and walked over to the pen, leaving him to join them up with the rest. The gate swung open and I stood well back, holding the end of the rope, while Garry brought the sheep to the mouth of the pen. Luck was with us and all but the last, a plain-headed ewe wearing a red collar, walked calmly inside. This ewe momentarily made a dash past the mouth of the pen, but Garry moved quickly, turning her in with the others. Much relieved, I closed the gate. (At the shed, the same plain-headed ewe obligingly paused and glanced over her shoulder. The chance was not to be missed. Garry darted in, causing her to go back on her haunches momentarily. She turned and, much to my relief, was easily 'worn'.)

The judges, Alan Gordon and Dave McTeir, shouted: 'That'll do'. The spectators clapped and cheered, I felt embarrassed, happy and proud of Garry, all at the same time. Our first run at a National was over. We left the field with a score of 182½. (We never did find out who deducted the half!)

Three weeks after the National, Garry took two first prizes in one weekend at Chatton Sandyfords, a trial which is held out on the heather, where a blustery

Above *Garry and the Coquet Shield.* **Right** *On the Whitelaw. Laddie (left) and Garry.*

wind blows continuously straight off the North Sea, and at Holystone, where he was presented with one of the most beautiful trophies on the trial circuit, the Coquet Shield, which we were allowed to keep for a year. This intricate shield, depicting a sheepdog, worked in silver and mounted on oak, dates back nearly 60 years. Inscribed on smaller shields around the perimeter are the winners' names. Many of them are well known handlers from the past hailing from both sides of the border.

Both these trials were won at the pen, Garry's penning tactics for once paying off. There had been numerous occasions before and since when they did not! On lighter sheep, his 'takeover bid' could have had different results. (It is said that the secret of consistent success is being able to read the sheep. Sadly, I am unable to do this. As far as I am concerned, trial sheep are as unpredictable as the weather and twice as windy!)

In September 1981, we travelled down to the famous three-day Longshaw event, reputed to be one of the oldest sheepdog trials in the country. The first contest had been held in 1898 on a day of howling gales and torrential rain.

Today this dramatic rock-strewn area, with its craggy, rugged beauty, is being preserved for future generations by the National Trust.

The Thursday and Friday trials are the qualifying heats for the exciting double outrun test which takes place on the Saturday. Local, novice and brace events are also included. Luck was with us and GWB, running Jed and Trim in the brace event, was placed third, and Garry, gaining the highest points on the Friday was presented with the Lonsdale Cup and given the opportunity to compete in Saturday's event with the chance of winning a valuable silver tea pot.

The sheep—white-faced Woodlands and crosses—couldn't have been better on the Saturday. Garry led the field until the last three competitors ousted him with brilliant performances. Harold Loates, with his faithful Jill, deservedly took first place, Clarence Storey with his prolific prize-winner, Blade, was second and Norman Darrel and Pat were a close third. Longshaw lives on in my memory as the Ascot of the trial circuit, but most of all it is remembered for the welcome we received and for the kindness of the people there.

On the Whitelaw
Garry set off wide up the south side of the steep Whitelaw, disturbing a covey of

partridges sheltering from the wind amongst the dying bracken fronds, away past the sweet scented spruce, on a left-hand gather, until he became a mere speck in the far-off distance. I remained motionless and quiet, watching for the first movement of sheep which would tell me he had arrived at his destination. Within minutes the first 'cut' began to move inwards with Garry following at a sensible distance until they reached a convenient position from where he could go back over the brow of the hill to gather up the late risers. This accomplished, they were brought gently, but firmly, downhill and joined up with the first cut. The ability of this curly-coated bundle of vitality, eyes ablaze with eagerness, never fails to fill me with a sense of humility.

He has an indefinable presence plus a natural 'bent' for putting large spread flocks together unaided, fetching them when possible at a steady pace, often through dense bracken and over soggy peat bogs. On these occasions I assist him with whistled directions to help guide him towards the many winding sheep trods made by countless generations as they wend their way at nightfall to the higher vantage points and during the daytime, to the sweeter, lower pastures.

Garry recast away to his left, answering my long, slow whistled command with alacrity, knowing the reason for it. The chill south-east wind carried my whistles speedily in his direction and caused me to shiver momentarily. Within minutes the three cuts were joined and scampering down the steep brae, rapidly in my direction. As they approached, their pace steadied until they slowed to a walk, and Garry dropped back. I could almost see what he was thinking: 'Is everything as it should be? Have I missed any sheep? Why doesn't she say something?' I reassured him with a 'Good boy!' and asked him to walk up. He happily obliged, depositing his charges at my, by now, frozen feet.

After a quick check to see that everything was in order we turned into the fierce wind and together drove the large flock of south country Cheviots down into the haugh where we left them in the shelter of a stone stell, to graze in peace until the evening when the hill tops would beckon them once again like some giant magnet, to the safety of their rock-strewn star-lit heights.

Chapter 7

The history of the Border collie's evolution

It is said that the wolf, the wild dog and the fox evolved from one common ancestor—a jackal-type creature. After domestication wild dogs became hunting dogs and herding dogs. It is most probable that they were divided originally into three distinct types—those which hunted with noses to the ground (which presumably evolved in dense undergrowth conditions), those which found game with heads held high as they lived in open ground and, last but not least, those which circled running birds and livestock. As time went by, man selected his hunting and herding dogs because of some favourable characteristic which made them superior to the others when it came to tackling both hunting and herding.

Eventually men hunted with nets. These were either set to trap game running into them or dragged over ground which was known to be a cover for birds and other small animals. It was most likely that through this method of hunting the noble pointer came into being. Man, the hunter, could not help but admire a dog which had a natural habit of slowing to a halt and remaining motionless at the correct distance when it scented game on the wind. This enabled the men using the dragnet to ascertain the exact position of their quarry.

The first pointers to arrive on our shores came from Spain and France. The advantage of using the dog from Spain was that he could work for long periods without seeking water. His only drawback was that, owing to his weight and size, he lacked the necessary speed and stamina needed for hunting and so both foxhound and greyhound blood was introduced. Once these crosses were accomplished, undesirable traits were bred out and the unique pointer characteristics retained and improved upon.

The Border collie has evolved from the wild, herding type of dog. When man domesticated livestock and wolves roamed the surrounding everglades, a large ferocious dog was required to keep predators at bay. The wild dog, long since domesticated, when crossed with the wolf (this particular beast being adept at circling its prey) would surely provide the ideal dog for this purpose. The wild dogs responsible for driving the prey must have been endowed with a high intelligence and determination to enable them to keep game on the move and in the right direction, often over difficult terrain.

Regarding method, the Border collie breed can be divided into two groups—the line (or driving) dog and the flanking dog. Presumably, prior to domestication, the line dog would have been responsible for driving the quarry into the waiting jaws of the flanking dog which had sped on ahead, as unobtrusively as possible, to lie in wait.

The dogs which carried out the actual slaughter were not necessarily the cleverest or the bravest in the pack. Generally, if a Border collie tears sheep he does so because he is afraid of them. Collies with courage will walk into sheep and move them without ever opening their mouths. There are occasions when sheep will attack the dog. In these circumstances the dog can be forgiven if he defends himself by giving the sheep a nip. I have seen good-natured collies merely click their teeth together as a warning. A genuine dog should be able to be termed 'soft mouthed'.

In the wild, pale coloured eyes are more prevalent in animals living in shady, wooded areas. Animals that have evolved with dark eyes deal better with bright sunlight in open country. Originally, body markings evolved as a form of camouflage so that the dog was able to blend in with the surroundings. Later man's preference for a certain colour meant that more animals of that colour would be bred.

The collie's white tail tip, inherited from the wild dog, has remained a dominant characteristic, making the breed, I believe, one of the oldest in existence. The tail tip is significant in that, like the fox, deer and rabbit disappearing at speed, it would signify danger and a warning to the rest of the pack and, in the case of young offspring, a desired direction of escape.

Glen and Laddie respectively, showing typical Pointer and Setter characteristics.

A versatile collie works anything that moves.

We have owned highly inbred collies which, apart from their black and white colouring, have resembled the fox in both looks and mannerisms. A red or chocolate-coloured dog of this type is often remarked upon as 'looking just like a fox' when running across a distant hillside.

I believe that the present-day Border collie's method of working, using 'eye' and style, is inherited from its sporting-bred ancestors, namely the setter, spaniel and pointer. 'Eye' and style invariably go together when present in the collie, however it is possible to breed a 'class' or quality dog occasionally which is relatively free of eye.

Centuries ago, in small villages, the in-breeding of different types of dog on a large scale was bound to have occurred and thus the dominant genes would quickly multiply. Imagine the sheer look of disbelief and amazement on the faces of the first people to have witnessed a dog of high quality going about his duties herding sheep, cattle, poultry and, on occasions, children within the settlement!

I will now describe the Border collie's sporting ancestors in detail.

The setter

In Great Britain the history of the work of setting dogs was first recorded by Dr John Caius in his book, *Englische Dogges* of 1576: 'These [setters] attend diligently upon their Master and frame their condition to such beckes, motions and

A true Pointer. Blackfield Gem, owned by Mrs Badenach Nicholson and winner of four Open Stakes and the Pointer and Setter Championship.

gestures as it shall please him to exhibit and make, either going forward, drawing backwarde inclining to the right hand, or yealding towards the left. (In making mencion of fowles my meaning is of the Partridge and the Quaile.) When he hath found the byrde, he keepeth sure and fast silence, he stayeth his steppes and wil proceede no further, and with a close, covert, watching eye, layeth his belly to the grounde, and so creepeth forwarde like a worme. When he approacheth neare to the place where the byrde is, he layes him downe, and with a marcke of his pawse, betrayeth "the place of the byrde's last abode", whereby it is supposed that this kind of dogge is called Index, Setter, being in deede a name most consonant and agreable to his quality. The place being knowne by the means of the dogge, the fowler immediately openeth and spreadeth his net, intending to take them, which being done, the dogge, at the accustomed becke or usuall signe of his Master, ryseth up by and by, and draweth nearer to the fowle that by his presence they might be the authors of their own in snaring, and be ready intangled in the prepared net.'

In early times spaniels and setters were often considered as one and the same, they were known collectively as 'sitters'. A painting by Albrecht Durer, who died in 1528, depicts a setter. He was a smaller dog, neater in appearance than his present day descendant. Other than this, the main difference was that, when standing to game, the 'Olde English' variety did so in a much more emphasised 'crouch', making the sharp knee bend of his raised foreleg even more exaggerated.

A century ago many setters failed to make the grade as far as field work was concerned. Some, in fact, just like some present day Border collies, appeared to be completely non-receptive and others were prone to bouts of running amok! However, as a loyal companion to man, a genuine setter is beyond comparison and is lovely to look at. It has classical lines, a gleaming, wavy coat and silken feathering on legs and tail.

The spaniel

Again it is said that Dr John Caius of Cambridge was the first person to describe 'Spaniells' in his *Englische Dogges*. Among the dogs used 'for fowling' he says: 'There is also at this date among us a new kinde of dogge brought out of France, and they be speckled all over with white and black, which mingled colours incline to a marbled blewe . . . the common sort of people call them by one general word, namely Spaniells, as though these kind of dogges came originally and first, all out of Spain.'

Long before Caius' time, the spaniel of Wales was referred to in the ancient laws of the country codified in the 10th century. It is described as being active and high spirited. Various encyclopaedias tell us that the French *epagneul* comes from *espagnol*, meaning Spanish or of Spain, but there is no real proof that the original spaniels came from that country.

One line of thought is that *epagneul* comes from an ancient French verb, *espanir*, which means to crouch or to flatten on the ground. The word 'spaniel' could also be a similar corruption of the Latin, *explanere*, which means to flatten out, and it is interesting to note that an old Italian verb, *spaniare*, means to get out of a trap or net.

In 16th century England there existed a semi-curly-coated water spaniel which

resembled a cross between a springer spaniel and an Irish water spaniel. This dog had an endearing quality—if his master got into any kind of trouble in mud or water, he would come to his aid, endeavouring to assist him by mouthing his sleeve. Some of my own collies show a similar trait in that if I become stuck whilst wading through a deep snow drift, the dogs come close in to my side until they feel my hand on their backs and then they move off steadily over the snow, pulling me out.

The water spaniel was said to be headstrong but more than compensated by being quick witted and thoroughly dependable. In the days of hawking, a leggy springer type of spaniel was used on the moor in rough heather. For working in dense undergrowth a short-legged, creeping spaniel was preferred, and so the cocker came into being—a cheerful, bustling, mud-loving character with nose to the ground, ideal for flushing out game.

Some people much prefer the way the cocker spaniel works, compared to the organised method of good pointers and setters. They get so absorbed with the job in hand that they become oblivious to everything else, but this can be very annoying at times, too. However, anyone with an ounce of humour cannot fail but be amused and cheered by such an animal who, on occasions, is capable of being extremely vocal and is quite a character too.

Spaniels originally were divided into springing and crouching strains, whence came the present day springer and the longer-bodied setters. The 4th Duke of Gordon bred the first Gordon setters in the early part of the 19th century. (These were out of a local shepherd's bitch called Maddie, a great pointer of game and got by a setter dog.) As there is often one or more red or deep orange whelps present in the best litters there is every likelihood that there is either 'Old English' or Irish blood somewhere in the line.

Gordon setters are capable of travelling without water much longer than most other sorts of setter and tend to show more variety in their attitude on 'the point'. Their tail (stern) can either be shorter than an English setter, or can be very long with a curl at the end and carried badly in action.

It is likely that many of our present day black and tan Border collies have Gordon blood in them.

The sporting dog will both scent and retrieve game, carrying it gently in his mouth whilst the herding dog is useful in that he can be sent over wide inaccessible areas to retrieve livestock on the hoof. The sporting dog makes full use of his nose to scent game; the herder uses his to locate sheep (sheltering from the hot sun and annoying flies) in dense bracken in the summer, or to find sheep buried in deep snow during the winter.

If, by chance, whilst out on the hill, a collie loses sight of his master (or mistress!) he will either raise or drop his head, immediately relying on his nose to find him. I have often observed my own dogs with some amusement as they zig-zag across some steep hillside hot on my scent, stopping occasionally to glance around, making full use of their remarkable vision.

Over the centuries Man has controlled the evolution of the various breeds of dog by in-breeding and cross-breeding them to suit his requirements. Thus the blood of dogs, both for hunting and pointing, greyhounds, herding dogs, setters and spaniels became intermingled. All would have been well if Man's nature was such that he could be satisfied with a 'useful' animal but, being of a competitive spirit, he required a 'superior' dog and proceeded to breed in earnest for such attributes as exaggerated style, speed and beauty, the latter practically bringing about the downfall of the pointer. English foxhound blood had been introduced initially to give the pointer more stamina, introduced again at a later date it was purely to improve him for the Victorian show bench!

Dr Caius remarked: 'We English men are marvellous greedy, gaping gluttons after novelties, and covetous cormorants of things that be seldom, rare, strange and hard to get'. According to an eminent Victorian writer, the 'hare foot' of the pointer was condemned and 'the hare brain was in ascendant'.

There is evidence to suggest that, in order to return the pointer to his former glory, some breeders introduced more greyhound blood into the breed until the only reliable strain emerged, that of the black or sable pointer, a dog frowned upon by show people because of his colour and type. In those days inferior working dogs were culled rigorously, so that, in the best kennels of black pointers only the dogs with the necessary attributes were bred from until, eventually, these dogs outshone by far their closest rivals.

The black pointer

After rummaging around in a second-hand bookshop in Berwick, I found a 50-year-old book about gun dogs and their training, written by Atwood Clark. This was my first introduction to the black pointer. The book contained a true account of one such dog who came into the hands of the author's father in the latter part of the 1870s.

It was intended that the self-coloured dog be used for breeding, and to quote the author, 'This dog was so near perfection that out of the people who saw him at work there was not one practical man who could find any fault at all in his behaviour, yet far less time than is usually devoted to dogs of that breed had been spent on his training.

'Until the last day of his life every gun before whom he ranged, either alone or with another dog, would closely watch the finished, faultless work of this unique artist and master of his craft.

'If his life history was written now in full, many incidents in it would no doubt be classed under the heading "Too good to be true", and few of those of us who knew that sable pointer are now alive to bear witness to his unique and subtly fine intelligence.'

This glowing description must fit at least one Border collie that each and every one of us has seen in our lifetime. My curiosity aroused, I set off in search of the black pointer, and subsequently found out the following:

A strain of fast, stylish pointers, some of which were black, existed as far back as the year 1644 and although black pointers are still seen in Europe, they are now believed extinct in Britain. The second Duke of Kingston owned a highly regarded kennel of pure black pointers. Each dog was sold for immense sums. (The first Duke had imported some liver and white pointers, the colours of the Royal French Kennel, in the year 1725.)

The stance and attitude of a dog tells a lot about his method.

At the beginning of the 19th century Mr W.R. Pape, a famous gunsmith living in Newcastle-upon-Tyne, started an excellent strain of all black pointers with a bitch from Admiral Mitford's famous Northumberland breed. Pape mated his bitch with a black dog he had imported from Spain, a powerful 'over heavy' dog, the wisest he had ever owned. Later Pape procured a strain of black pointers with white between their forelegs from a Mr Usher, and there is little doubt that Pape used greyhound blood in improving his dogs. Towards the latter part of the century an 'improved breed' of black pointers purchased from Mr Pape and re-christened 'Arkwright Pointers', gained popularity in the hands of William Arkwright, a well known dog breeder and judge.

In the middle of the 19th century Lord Home possessed two seperate stocks of black pointers—one at Douglas Castle, Lanark, the other at the Hirsel, Coldstream, in the Scottish Borders. In 1842 the Douglas Castle pointers were not entirely black, but eventually became so. The late head-keeper at Douglas Mr Amos, wrote of 'Sweep', a first rate dog which Lord Home lent to William Arkwright in 1895, that he (Sweep) was out of a bitch of their own whose ancestors were 'all jet black', very handsome and good.

Mr W. McCall of Ferniehurst, Jedburgh, head-keeper of Lord Lothian, wrote (in January 1902) that he first got his noted strain of black pointers from a keeper on the neighbouring estate of Hartrigg in 1864, and had the breed in his possession for 36 years. W. McCall's son, William, keeper at Glen of Rothes, was also 'possessed of it' and, at different times, he used black dogs of Lord Home from the Hirsel, of Sir William Elliots from Stobbs Castle, Hawick, of Lord John Scotts from Spottis Wood, Lauder, and of Sheriff Rutherfords from Edgerston,

Jedburgh: McCall Junior said that, for many years, he bred his dogs with the Douglas Castle dogs, and that he first saw black pointers in 1852, when he went as under-keeper to Drumlanrig Castle, which belongs to the Duke of Buccleugh.

The black pointers tended to be smaller in stature than English pointers, they had a pronounced 'stop' where the nose juts out from the face, hare feet (necessary attributes in a dog that has to make quick turns and sudden stops!) and a black mouth and lips—much valued traits in today's Border collies.

The Victorians were notorious for their experimentation and interest in improving the appearance of breeds, purely for showing, and so, much to the detriment of the pied English pointers, harmful cross-breeding took place and hound blood was infused in the strain.

In 1906, after nine years of research, William Arkwright published the second edition of his immortal classic *The Pointer and His Predecessors,* and proved without a doubt, that he was a breeder who had the interest of the working dog at heart. It was this genuine belief which eventually led to his resignation from the Kennel Club. I found a passage in *The Pointer and His Predecessors* on the pointer which may equally apply to the collie today: 'It is just this perfection that I have now to describe and consider—a perfection that required a hundred years of practical work, attention, and combination to mature. For though the pointing dog first touched the shores of England in the beginning of the eighteenth century, the pointer did not reach his Zenith til much later. This type unfortunately is now rare; because fine old races of animals, unlike old wines and furniture, do not placidly await the re-awakening of good taste, but during periods of neglect tend to disappear and die out. Fortunately, however, the true pointer is not quite extinct, though he is nowadays to be found more often in the Highways and Byways than with number and pedigree attached to his name.'

The black Arkwright Pointers flourished and grew in popularity—they were excellent workers in the field and were noted for their outstanding style and stamina. Their superior intelligence ensured that they required very little training. In short these were 'natural' dogs with an above average ability for winding game.

In the 19th century the fast, stylish pointer was much more highly regarded than he is today, his most ardent admirers being wealthy landowners. When it came to hunting, finding and pointing game the black pointer was beyond comparison, especially when working on Northern hills and in rough heather. In those days he represented the very best of the breed with bloodlines which were a closely guarded secret. Only the most attractive, intelligent and stylish animals were used, much to the envy of friends and neighbours. (A man was judged by the dogs he owned.)

★   ★   ★

One of the nicest book's that I have read is *Dogs: Their Points, Whims, Instincts, and Peculiarities,* edited by Henry Webb and published by Dean & Son in the 1870s. The following is an exact reproduction of Mr W. White's excellent contributed chapter on the 'Scotch Colley'.

The Scotch Colley

By Mr W. White, of North Farm Kennels, Bleasby, Southwell, Notts.

The Scotch Colley, to my thinking, when well bred, nicely proportioned, and pleasingly coloured, is perhaps one of the most handsome of all our British dogs. With beauty he combines goodness, *i.e.*, working powers, and is usually possessed of an almost iron constitution, which seems as it were regardless alike of summer suns or winter snows. He is seen toiling day by day after his charge with untiring energy, for which this breed of dogs are very remarkable; and this feature in them is the more astonishing when we so often see them kept by cruel masters in a state of semi-starvation, ill-fed and little cared for, left outside the shepherd's cottage door for the night, midst winter's piercing winds and falling snows. Nature here, as in all her bounteous and wondrous works, has protected the colley with a coat so thoroughly anti-rheumatic in character and entirely impervious to sleet and rain, that he may be seen lying about on the ground whilst tending his flock during the most inclement weather, with, as it would almost appear, some degree of pleasure to himself.

Another grand feature in his character is the great love he has for and the strong attachment he forms for his master, the summit of his happiness is being in his master's presence—

> *'The straw-thatched cottage, or the desert air,*
> *To him's a palace if his master's there.'*

I am not aware of any class of dogs that so quickly forgive and forget the rough usages of their masters as colleys do; and it is perhaps a blessing they do, for I fear a great many shepherds treat their dogs in a very unmerciful manner at times, and that for little or no fault on the part of their sagacious dog, but from inaptness on their own part to convey to the dog, by word or signal, really what they require from him. It is not to be expected that dogs can be taught to understand half the words in *Walker's Dictionary*; hence the folly of using numerous words— the fewer and more expressive they are the better—always using the same words for the same requirements. With colleys, as with sporting dogs, signals may be used with advantage. The well-bred Scotch colley differs to some extent from our English sheep-dog: in point of size he is not so large; the ears are smaller; eyes rather small, keen, and with some degree of fire in them; the muzzle finer; skull narrower and more foxlike; mask smooth, with a coat of woolly hair shorter in the body-coat, and standing off more than in the English dog; the neck is well clothed with hair, much longer than his body-coat, and is usually, by admirers of the colley, called his mane; which, if he possesses somewhat in abundance, imparts to the dog a very noble and striking appearance; the hindlegs are straighter than in the pointer or setter, and smooth from the hocks downwards; the feet should be round and compact; the loins strong, and back ribs deep, tail bushy and some-what symmetrically carried, the tip being carried up and over its own root (for show purposes), and quite free from any approach to corkscrew curl. Perhaps as

Cockie, the Scotch Colley.

good a specimen as it has been the lot of most to see (for all points) is the well-known champion Scotch colley dog Cockie, whose portrait, a very truthful one, is annexed. In this opinion I am borne out by most of our most eminent judges. Cockie has contested at most of our largest and best shows of dogs, and is the winner of a number of money prizes and two silver cups. To mention a few of what I consider his greatest successes may not, on passing, prove uninteresting to the reader. He competed at the Border Counties Show at Carlisle, for a champion colley cup (when an open one), presented by J. Wright, Esq, for the best colley of all classes; which he won. He also won their first prize for rough-coated colleys, beating a large number of others; the judging of the colleys lasting some two hours and thirty-five minutes. I afterwards heard a great portion of the dogs were from the mountainous districts. I may say I felt a desire to see dogs from Scotland, and to compete with them; although this was not a very agreeable task, owing to the distance from where I reside, and being held in the latter end of January, 1870; and I well remember the weather not being sultry, but, as Paddy would say, 'quite the reverse.' *Nil sine labore.* Quite satisfied, I placed Cockie in his hamper and started for home late on Saturday night, which I reached on Sunday morning, at 10.30, travelling the night through. Nothing daunted, on the following Monday I started with Cockie for Maidstone (in Kent) Show, where I was rewarded by his winning the Mayor of Maidstone's cup. He competed also at Darlington, in Durham, a great sheep-dog district, where he won first in a class numbering thirty-four; here I was pleased to see a kind of platform raised to stand the dogs on, any malformation being easily discernible by this means. Cockie was placed second at Birmingham Show, in 1870; second at Birmingham in 1871; and first at Birmingham, 1872, their last show; and has also been a winner at Birkenhead, Hanley, Middleton, Newark, Islington, Wellington, Boston, Gainsborough, Nottingham, and other shows; and has beaten, at the time I write this, every dog against which he has competed. In colour his body-coat is something like seal-skin, blackish outwardly, and when parted by the wind, showing a rich fawn colour beneath; for body-coat I prefer this colour, because it is only found in this class of dog; black and tan is also a fashionable colour, but much as I admire it in the English terrier and setter, I don't care for it in colleys; some admire a white line down the centre of the face; I don't; I prefer a white collar and chest and forelegs, with tag on tail, as giving more the character of colley to the dog and more room for a picture worthy of Landseer or Ansdell. This dog is very affectionate and greatly attached to his master. My kennel boy, Charlie, said one day, 'Why, sir, if Cockie could only get out the first word, I'm sure he would talk to you.'

When used as a companion at night, he is one of the best. Should the road be somewhat lonely, he walks behind; if any person approaches (long before I can either see or hear them), he utters a low growl, which grows louder and louder as they near me; if he considers them suspicious characters, he would stop them were I not to restrain his anger: he being very apprehensive of danger at night time, the darker the night the more watchful and determined he seems to protect me, so much so that I usually place him in a collar and lead-strap when darkness comes on; in the day time he has a wag of his tail for most he meets. He is very fond of a gun, and is an excellent water-dog in all weathers. The points of the colley I would place as follows:

Head, ears and eyes	25
Colour and coat	25
Chest	15
Legs, feet, and shoulders	15
Back and loins	10
Tail	5
Size	5

In these points you will observe I differ with the points of the National Dog Club on sheep dogs, as I give coat 25 with colour; for without the proper texture of coat I count him next to useless in standing the weather. Chest I give 15, as speed is very desirable for heading, be the dog what he may in other respects. Dew claws I have not mentioned, as I don't consider them at all indicative of good breeding, every litter showing examples with and without.

I write the above, being strongly pressed to do so, and it is my private opinion with reference to the Scotch colley, but I won't say I am right.

The editor of the *Banffshire Journal* gives us the following anecdote, the truth of which he can vouch for: Mr Shaw, Achgourish, Kincardine, Abernethy, with his favourite dog Chance, left home for the hill, for the purpose of what is called 'the sheep gathering'—that is, bringing them down to a convenient place to be shorn and washed. They had not proceeded far, when Mr Shaw, from indisposition, or some other cause, did not feel inclined to go up the glen, and he told his dog to go away and bring down all the sheep, and that he would await his return. Chance instantly obeyed his master's order, went up the glen, gathered all the sheep together, and came away with them exactly in the direction of his master. We may mention that Chance's movements were observed from the top of Craigourie by the hill pundler. Mr Shaw, who waited patiently the return of his faithful servant, now saw the sheep nearing him to the west of Craigourie, and at this moment observed a hare getting up amongst them, and looking very bewildered. Chance, taking opportunity of this, left his charge for a little, and took to the chase, and, after some stiff work, succeeded in catching the hare. Mr Shaw called out to the pundler to go and take the hare from the dog. Chance, anticipating what was to follow, surveyed with suspicion the pundler, who was fast approaching him. Yet not liking to do battle with one with whom he was on intimate terms, instantly threw the hare over his back, as being the easiest mode of carrying, brought with him the sheep with all speed, and laid the hare at his master's feet. Recently, the same dog was asked by Mr Shaw to go and keep the crows out of the potato field. This he did, and in about half an hour returned to the house with a live crow. It is supposed he concealed himself below the stems, and when a flock of them alighted, made a dart at them, and in this way had caught it.

★   ★   ★

The first dog show to include the working collie was held at Newcastle-upon-Tyne in 1860 when five collies represented their breed. By 1885 over one hundred show collies were competing against each other in various classes at the Crystal Palace Royal Aquarium. The collie was always a firm favourite with Queen Victoria who kept them over many years and at one time she had a rare white collie in her possession.

In 1876 a sheepdog trial was organised and held at Byrness in Northumberland, a stone's throw from the Scottish Border. The trial, one of the first in the area, was won by Walter Telfer. Five years later it was the turn of Adam Telfer, to take the honours and come away with the much coveted first prize. It is perhaps worth a mention that these were the men responsible for providing much of the foundation stock from which the present day show collie evolved. In all probability this foundation stock, although well bred, would have consisted of 'whites' and non-workers of various hues.

Then, suddenly, towards the end of the 19th century, like a bolt out of the blue, there appeared on the sheepdog scene a dog of great class, character and genius, the likes of which had never been seen before, nor was likely to be seen again. He made such an impact that it is said that those fortunate enough to witness him at work never forgot his style and uncanny ability, and that his sheer excellence was imprinted in their minds forever. For this dog was Old Hemp, procreator of our present day Border collie. His master was Adam Telfer Sr who said of him: 'Hemp flashed like a meteor across the sheepdog horizon, there never was such an outstanding personality.'

Hemp's description is as follows: 'He was a powerful looking dog, sturdy, short coupled, and he had a gently curving spine. His head was masculine and strong with a dominant expression and very pronounced 'stop'. His muzzle was deep and his ears wide, set low hanging. His coat was the colour of jet, and there was very little white about him.'

Hemp's sire was the easily handled and amenable Roy. A plain worker, black, white and tan in colour, whilst Meg, Hemp's dam, was entirely different. On one side she was bred from the Old Rookin Whites who lived at Rochester, near the Scottish Border. She was of a very high class—extremely stylish and strong of eye and noted for her intelligence and extreme sensitivity. Furthermore Meg was self-coloured—as black as the night sky.

It was around this period that a great improvement became noticeable in the average working collie. The bark and bite necessary in a rough herding dog began to disappear and the tail which, prior to that period, often resembled either a corkscrew or a flag pole, if old engravings and paintings are anything to go by, became well set on and more settled.

The International Sheep Dog Society was formed when Scottish and English sheep men met at Haddington East, near Edinburgh, in 1906, however it was not until 1955 that the then secretary, Mr T.H. Halsall, amassed incomplete sets of registration forms and built up a card index of more than 14,000 entries to form a stud book, with Telfer's Old Hemp at number nine considered by everyone to be the vital cog.

Laddie taking his daily dip.

The intention of the founder members of the ISDS was to stimulate interest in the shepherd and his calling, and to secure better management of stock by improving the shepherd's dog. Without doubt interest in the shepherd has been stimulated far beyond expectation, however, whether or not the shepherd's dog continues to improve is (according to some flockmasters) a debatable point. In years gone by our forebears went to great lengths to breed quality into their hunting 'sporting' dogs. Nowadays a collie with real quality is rarely encountered. Quality has been lost through in-breeding, and inferior animals being bred from. It is seldom realised by dog breeders that, in multiplying the good genes, they are also multiplying the bad ones and encouraging 'throw backs' to the wild dog, the wolf and fox.

Too much collie in-breeding increases the resemblance to early ancestors, although not all throwbacks are necessarily bad. For example, pricked ears often mean improved hearing and so can be regarded as a good throwback. On the other hand the killer instinct, which still causes a dog to use its teeth at the slightest provocation, and dishonesty, whereby a dog cannot be trusted out of sight, are examples of bad throwbacks.

Close in-breeding with faulty bloodlines can also bring about hereditary defects such as blindness, deafness, hip dysplacia, lack of stamina and numerous deformities plus an excitable temperament.

In addition, close in-breeding between Border collies has produced some interesting results as well. Colouring and appearance can vary enormously—besides the normal, cleanly-marked black and white dog (preferred by most) there are chocolate, yellow, blue, white and mottled varieties. There are short-coupled dogs and dogs which are long in the back, giving the appearance of possessing an extra rib. There are long-legged types, short-legged types, dogs with round strong feet and dogs with narrow hare feet, dogs with pricked ears, wide set ears, long spaniel-looking ears and short ears. Some dogs have strong, deep muzzles or narrow 'snipe' noses, half-white faces or all-white heads which often show the characteristic 'spaniel spot', between the ears.

A dog's habits and abilities are often affected by in-breeding. Some dogs have a working method whereby they weave backwards and forwards mindlessly, like a spaniel quartering game. You find dogs which can line sheep up, and dogs which cannot, dogs which work with heads held high, and dogs which work with noses to the ground, dogs with natural distance when working sheep, and dogs which work too close to them. Some collies' idea of heaven is to wallow in the nearest mud hole, whilst others are scrupulous to the extent that they would rather die than look dirty, insisting on a daily bath in the nearby burn, swimming in circles and playing like otters, much to my amusement. Not surprisingly the dog who is particular about its appearance tends naturally to be clean in the kennel, unlike his mud-wallowing brother who will think nothing of fouling his kennel and then sitting in it! On occasion a foxy type dog appears which is prick-eared, light of frame and very pretty worker that catches the eye at first but does not hold it for long.

In years gone by there were many strains of working sheepdog which were similar in type and ability. Owing to the quality and style of the collie in the Scottish Borders (which were handled skilfully by the many competent Border handlers of the late 19th and early 20th century) many of these strains no longer exist today.

In those days, the hill shepherd was his own master and, to a certain extent, he still enjoys that privilege today. Sheep destined to be sold were shed off on the hill and driven to the market by the shepherd and his dogs. (Some of the old drove roads can still be seen today.)

The Border handler is still revered for his ability at sheepdog trials, and his shepherding skills, both on the hill and in-bye. Border sheepdog trials reached a very high standard in the past. Sadly, that is not always the case today. The sheep's behaviour has deteriorated, mainly because the average shepherd is now required to look after two hills, or if in-bye, twice as many sheep, with the result that the sheep are handled less and therefore less amenable.

With the deterioration of the sheep's behaviour comes a deterioration in the dogs required to herd them. It is nigh impossible, when competing in sheepdog trials, to have a good run on wild unmanageable sheep when using a powerful dog. It is because of this that our average present day collie tends to be easy to command and is pliable, and lacking in initiative and determination.

Up and down the country a large number of dogs of this calibre are being used at stud, purely for the purpose of breeding potential trial winners. When the name of a particular dog appears half a dozen times in a pedigree it is more than enough. 'In and in breeding', as the saying goes, inevitably leads to 'out and out FOLLY'. Without any doubt, the sheepdog of yesteryear was certainly more intelligent, determined and capable. He had to be in order to do what was expected of him.

I fear that the greater the popularity of trialling, the more attributes the dog will lose. The hill man's requirements in a dog, namely power and practical ability, will always be the same and so it is to these men and their dogs that sheepdog trial enthusiasts will have to look to get the Border collie back on its former firm footing. To keep him there, more hill trials with long gathers and more double outrun trials will need to be organised. An absence of 'fetch gates' at trials so that the dog could be left to fetch sheep on his own plus a deduction of points for over-commanding would greatly benefit the breed.

Chapter 8

The attributes of a good Border collie

The tail

The reason why I prefer a sensitive dog is because I find that type to be much more receptive than their thicker skinned brothers. A dog does not have to be bold with strangers as long as he works with a purpose and has no fear of sheep. When a mature dog is working, he should carry a tight tail which gives the appearance of being glued to his hocks. There is a great deal of truth in many of the old sayings, especially the adage: 'If the dogs tail is right, his head is!' and it is possible to breed a collie which does not throw its tail when putting in a tight turn, though admittedly they are few and far between.

Some young dogs, on first sighting sheep, lift their tails slightly, rather like the pointer and setter. This habit is often a sign of intelligence, and the dog does it when he is weighing up the situation and anticipating the next move. By the time he reaches his second birthday and is beginning to know the ropes, his tail will have settled.

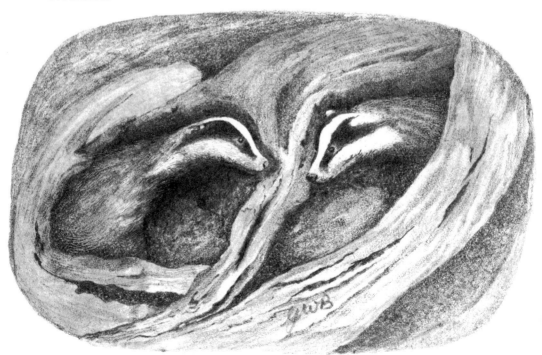

The more particular the breeder is to eradicate glaring faults, the better his litters will eventually be. Instead of there being just one decent pup in a litter, it will be possible eventually to breed litters of good pups. Line breeding is rewarding when there are generations of powerful, sound hill workers a few generations back in the pedigree. We have bred better litters this way than close in-breeding or by using an out-cross of completely different blood.

A calm, intelligent dog carries a loose low tail which tightens and curls into a classical walking stick shape when he 'Warms up' on sheep. This type of tail is a combination of beauty and perfection. The tail which is carried between the dog's legs and sometimes along the belly can denote a timid nature. This type of dog benefits from continual reassurance to boost his ego. Regular work amongst other shepherds and their dogs will help him gain confidence. Keeping this kind of dog on a chain does him more harm than good as he is bound to feel trapped and therefore frightened.

Young impetuous dogs often flap their tails over their backs when asked to put in a tight turn on sheep. With quiet and careful handling, most dogs grow out of this awful habit, although others through bad breeding do not. Impatient, excitable handlers are often the cause of the tail flip in a mature dog. It is a sure sign that the dog is confused, either by his handler or the situation he is dealing with. Old dogs lacking in speed can be forgiven for taking short cuts, and for using their tails to act as a brake!

It is obvious to all that a happy natured dog will wag his tail at every given opportunity. It is always pleasant to be greeted in this way, however, an over friendly dog does not always take life or work as seriously as he should. Some dogs take life extremely seriously. Their lack of obvious frivolity is, I think, a good sign and can be soothing to someone in a serious frame of mind. This type of dog is more introvert and rarely goes 'where angels fear to tread'! or makes a nuisance of himself. The only occasion when you will glimpse him delirious with delight, is when he has been deprived of seeing you, or sheep, for a few days!

After the head, the tail is a dog's most vital piece of equipment. Not only does it tell us how he feels, but it shows his reaction to practically every circumstance he encounters. Watch your dog's tail and you will know him well.

Some dogs, often the inbred sort, chew their tails, giving them a ragged appearance. They do this either because they have full anal glands or because they are just bored. Occasionally a dog's tail will twist right or left, giving an unbalanced appearance. This fault is simple to correct and is best done when the dog is young. Whilst the puppy is eating his supper, gently manipulate his tail, using thumb and forefinger of each hand until it is straight. You will probably need to correct the tail on more than one occasion. It is a job that is well worth the effort as nothing looks worse than a tail that is not set on straight, and nothing looks better than a tail that is!

The eye
There is a great deal of truth in the saying that the eyes are the windows of the soul. I prefer to avoid dogs with black eyes which resemble fathomless pools, as it

Above and right *Jan working sheep and showing the 'eye' of a good dog.* **Below right** *Laddie diplaying my favourite colour combination in a dog.*

is impossible to see any expression or to know what they are thinking. In most cases, I have also found them to be of low intelligence. Pale-eyed dogs are supposed to be cowardly. I have known many courageous pale-eyed dogs but they did tend to be cunning and occasionally lazy. My favourite kind of eye is a rich red-brown colour, followed closely by a nut brown. Large eyes, full of expression, are to be preferred to small, 'piggy' eyes which give dogs a mean look.

When one talks of the 'eye' of a dog one is really refering to the hypnotic stare which the dog fixes upon the sheep. I have seen a determined 'strong-eyed' dog come to a complete halt when attempting to push slow heavy sheep round a trial course. The dog gets to the point of balance and holds the sheep with his 'eye'. The dog refuses to move, as do the sheep and the result is stalemate! The late Mr C.J.Cornish gives the following definition on the habit of pointing in sporting dogs. It could easily refer to what sheepdog handlers call the 'eye': 'The pointing habit is an artificial prolongation of the instinctive pause of a carnivorous animal on first becoming aware of the proximity of prey—a pause for the purpose of devising a stratagem. Indeed it is a quality so readily intensified by selection that in specialised breeds of dogs there is a constant danger of its exaggeration, evidenced in false pointing and unwillingness to road up the game, both of which habits maybe reckoned among the mortal sins of a gun dog.' Thus 'eye' in a sheepdog is important for it aids him in the control of sheep.

Coat and colouring

Hill shepherds often show preference for a smooth-coated or bare-skinned dog. The obvious advantage is that these coats do not 'ball up' or become weighed down with ice and snow. The disadvantage is that during prolonged hard weather, dogs do feel the cold!

A well formed, strongly made, bare-skinned dog is a more attractive proposition than a whippet type of poor conformation.

In the old days the drovers were often accompanied by a fast hunting dog to ensure that, come nightfall, the cooking pot was never empty. Our present-day smooth-coated collie has more than a dash of greyhound blood in his veins and is quite capable of coursing and catching a hare. It is obviously fleeter of foot and better at turning sharply than its average rough-coated brother and is often endowed with more stamina due to less in-breeding. (I admit to showing a preference for a long-haired dog, purely because of its appearance and I am quite prepared to thaw him out in the kitchen should the need arise.)

I was curious to know how a long-coated dog fared in a hot country and was assured by a regular overseas visitor that a long coat wafting up and down as the dog ran, created enough draught to keep him cool. The same man, to prove his point, wears trousers and a jacket in a temperature of 100 degrees, as he says: 'What keeps the heat in, keeps the heat out!'

A very black dog can have the effect of startling sheep, whereas a white dog can produce the opposite reaction if it is weak or steady going. Generally speaking the lighter coloured the dog is, the softer he tends to be. A white dog which is afraid of sheep is of no practical use and can end up being chased out of the field by the stock it is working. A reasonably powerful dog with a dark body and a fair bit of white about its head tends to have a calming effect on sheep. Therefore my favourite combination is a dark-bodied, powerful dog with a full white brow.

The size

It is worth bearing in mind that small and medium sized dogs are often hardier and last longer on the hill than large dogs. They are also fleeter of foot and more nimble at putting in a tight turn, especially when penning awkward sheep. A large dog, as he ages, often gets heavy and slow and is more likely to suffer heart strain than his lighter framed brother.

Chapter 9

The mating game

'Working dogs are, of all the domestic animals, the most interesting to breed, as they alone, in addition to the physical, have mental qualities that require solicitude—William Arkwright, 1906.

Dogs vary enormously in regard to sexual drive and prowess. Some dogs exhaust themselves mentally, howling constantly throughout the night and vanish into thin air the moment you open their kennel door, after just the suspicion of a whiff of a 'hot bitch' in the area. At dipping time and when you have a hill to gather this type is next to useless. Many a good dog is ruined by his owner allowing too many bitches to come to him.

Mating in abundance is bound to be energy sapping and can definitely make a dog go 'off the boil' as far as giving a polished performance on a trial field is concerned, especially in the case of the least determined, easily handled sort of dog. On rare occasions you will meet up with a dog which is exactly the opposite and can 'take it or leave it' when it comes to sex. My dog, Garry, comes under this category. When given the choice between mating and work, he will choose work every time and has been known to work beside a 'hot bitch' all day ignoring her overtures because he was simply not in the mood.

In his youth this attitude proved embarrassing to us and frustrating for the bitches who were brought to him. After one occasion where he retreated to contemplate a cobweb in the grimy lime-coated interior of our byre rather than do what should have come naturally, we finally got the message, coming to the conclusion that as a 'stud' our Garry was a complete washout! Reluctantly, we decided not to take any more bitches to him unless we wanted a pup for ourselves.

Thankfully there is an exception to every rule. In Garry's case, this exception came in the shape of a sweet, pretty little tri-coloured bitch called Nell (Leadburn Nell to be precise!) Garry's protest against the energy-sapping rigours of mating came to a sudden and abrupt ending, where Leadburn Nell was concerned. He was well and truly smitten! Could it be her prefix? we asked ourselves. Was Garry a snob? Leadburn Nell belongs to William Black of Bowsden and I consider William and his mother, along with Sarah Morrison of Cleugh Brae, to be some of the best rearers of puppies in the Border Country!

After suitable periods of courtship and lengthy overtures, Leadburn Nell, despite a horrific accident when she was mown down by a van and had to be completely 'rewired' internally, has successfully bred and reared five nicely marked litters of very intelligent puppies by Garry. From each litter we have taken a 'service pup' and have been well rewarded for Garry's efforts.

Garry, usually a loner, goes through a complete change of character whenever a new young offspring appears on the scene. It is as though a dormant instinct to protect makes its way to the surface and takes over completely. Normally walking on in front so as not to be forcibly involved in the boisterous ear nipping and tail pulling games of his older sons, he now 'hangs close' keeping a watchful and protective eye on the puppy who is usually totally unappreciative of his efforts. If Garry considers the puppy to be in any way threatened by the 'big dogs' he intervenes by fixing the offender with a fierce glare whilst uttering a quiet growl, this usually being all that is required to send the offender on his way.

*   *   *

As most bitches become mothers, they tend to mature mentally at an earlier age than a dog, are quicker to start work and are more devoted to their owners, although some dogs are extremely faithful too.

When a young bitch is eating her food, she is often more inclined than a dog to bite her owner or his children should they approach her, and therefore must be treated with respect at meal times.

For practical work, a masculine or 'doggie' type of bitch is preferred to a very feminine one. Unfortunately the former tends not to be very fertile and usually produces small litters, if she breeds at all. Quality and quantity infrequently go together.

If you already own a useful, well bred bitch you have a good chance of starting your own 'line'. If you are a member of the International Sheep Dog Society you may wish to add a prefix to the name of her offspring. This can be of your own choice (providing it has not been taken out by somebody else). When searching for a 'would-be' mate for your bitch it is advisable to look for one which has the attributes which she lacks.

On average, a bitch comes into season twice a year. There are some that only come 'in' once, and others (thankfully rare) that are in season every four months. Some appear never to come in season or do so so slightly that they can go undetected. Sometimes a change of environment is all that is needed to induce a bitch to come 'full on'.

At the onset of heat, the bitch becomes very friendly. Her vulva become enlarged and blood is passed. Some bitches will take a dog at any time during their heat and produce puppies. Generally a bitch's fertility is at its peak towards the latter part of the heat, when the bleeding has become brown and watery. She should be mated on more than one occasion to be sure she becomes pregnant.

There is no definite length of time that heat will last, most bitches vary, also age must be taken into consideration. Usually when the bitch will stand for the dog, she is ready to be mated. Often 'heat' will go off a bitch after she has been mated with success.

Mating is not always as straightforward as it would seem. It is sometimes beneficial to apply Vaseline to the bitch's vulva. A bad tempered or nervous bitch which may be inclined to attack or worry her would-be mate should be firmly held, or the dog will run the risk of having his ear chewed off! A friendly natured bitch can be allowed to run free for a while but, once the dog enters her, she must be held and his hind leg must be lifted carefully over her back so that the dog and bitch are facing in opposite directions. (Dogs have been known to injure themselves when not enough care is taken.) If a rough-coated bitch should prove difficult to mate, she will benefit from having some of the long hair around her vulva removed. A small pair of scissors can be used for this purpose.

Garry keeping a watchful eye on Glen.

The paternal Garry.

A bitch will carry her puppies inside her for approximately nine weeks, give or take a day or two. During her pregnancy it is important that she receives a balanced diet which includes, whenever possible, meat, fish, milk and eggs. Half-way through her gestation, the bitch must be wormed. This is very important to ensure that both she and her puppies, when they are born, will arrive on the scene as worm-free as possible. When the bitch is not wormed, her puppies are born with them, and consequently suffer, especially just after a meal when they will be heard to whimper pitifully. Not all pups pass worms in their excreta; others do and, when severely infested they will also vomit them.

The spring, for obvious reasons, is the best time for puppies to be born. If the weather is not good then an infra-red lamp should be installed in the place of whelping, at the correct height so as not to cause danger or discomfort to the new mother and her offspring.

Puppies can be born quite safely on a clean wooden floor. If straw is used for bedding, they can become entangled in it. Sawdust is inadvisable as it gets up their noses and into their mouths.

Occasionally a very young or an old bitch will have difficulty delivering a large puppy. Most shepherds are skilled in midwifery but, if you are in any doubt, you should consult a veterinary surgeon so as to alleviate any risk.

The 'nest' must be spacious enough for the bitch to turn round in, but not so large that the puppy can crawl away from its mother's warmth and perish from the cold. If by chance a mishap should occur and a puppy does become chilled, it is best to wrap it in tin foil and thaw it out gradually.

After a bitch has finished whelping she should be given a small drink of warmed milk. (She should have no solids until the following day so that her stomach can settle after the upheaval of giving birth.) Milk is very good for the bitch in that it provides her with necessary calcium. A bitch which does not receive sufficient calcium can go down with milk fever. The symptoms are convulsions followed, if left untreated, by death. Calcium deficiency is more prevalent in the young bitch who possibly finds it more difficult to assimilate.

Clean water in a shallow dish must always be available and put in a safe place so that the puppies cannot fall in and drown. If the newborn puppies look empty and are constantly whimpering, the bitch is most probably short of milk. Puppies can easily be topped up with a proprietary milk substitute specially made for them, using a syringe, or special puppy bottle. Take great care not to 'drown them with kindness'. In order to 'do' her puppies well, and at the same time keep the flesh on her back, a bitch who is suckling puppies needs as much food as she can eat. 'Little and often' is the rule.

It is wise to keep young children away from a new mother unless they are known to her and she has a particularly gentle disposition. I have seen bitches leap out of their kennel and attack a child, believing that it meant to harm their offspring. A very young bitch's puppies should not be handled until a couple of days after their birth as I have known them to eat their offspring owing to the smell of human hands.

Chapter 10

Puppies

Three-day-old pups should have the insides of their hind legs examined for dew claws. These are easy to remove with the aid of a helper and a pair of small scissors. The minute wound should be dabbed with disinfectant. Once the puppies become active, they can be bedded with a little clean straw and regularly cleaned out so as not to allow the kennel to become fouled, thus providing excellent conditions for vermin to breed in. Most pups open their eyes by the 12th day. Any with 'gummy' eyes should have them bathed gently with warm tea. (Do not ask me why but cold or warm tea is a good old fashioned remedy for bathing eyes!)

Three weeks is a suitable age to dose a pup for round worms, followed by another dose ten days later. Any puppies kept after three months will need a repeat dosage, increased appropriately according to their extra body weight. If you should by chance purchase a pup of this age and you suspect him of being worm-ridden, you will need to relieve him of them gradually. Start off by giving him half the normal dosage of pills, otherwise a mass of dead entangled worms (closely resembling bamboo shoots!) can cause an internal blockage. A pup with this problem looks very sorry for himself indeed. If he tries to eat, which is doubtful, he will probably vomit. An operation is usually necessary for a bad case and, provided the puppy has not been left too long with worms, he will make a complete recovery. Puppies sometimes need worming for tape worms, too. These are the flat, cream-coloured segments which either adhere to the hair beneath the dog's tail or come out in their excreta. En route to a sheepdog trial one of our bitches, much to our surprise, grandmother's horror and junior's curiosity, managed to regurgitate a giant tape worm!

At around three weeks of age you can begin to introduce the puppies to a little warm milk by gently dipping their noses and fore paws into it. From milk they will soon progress to three meals per day of puppy meal—milk, mince and gravy and be weaned off the bitch gradually until, at around seven weeks of age, they are allowed to go to their new homes.

Always remember to tell the new owner exactly what the puppy has been fed on, that he has been wormed but will need worming again at three months of age, and that he will require some vaccinations then, too. Try to ensure that the poor little mite goes to a kind home where he will be well fed, well housed and properly treated; it is the least you can do considering that you are responsible for his birth.

Any puppy in the litter failing to come when called or refusing to awake when you clap your hands is probably deaf, an ailment for which there is no cure. It is a genetic problem appearing in white or mostly white puppies. I recently read about a deaf collie bitch who had been taught to work by hand signals. I admire her owner for his great patience and obvious kindness.

Should it be necessary, for any reason, to have a dog put down, it is more humane to allow a vet to do it. There are, sadly, a few shepherds who still hang or drown puppies, possibly because their fathers, and their fathers before them, put their puppies down that way.

It is necessary to supervise dogs at feeding time when more than one is kept in the same kennel, otherwise you will end up with a situation where 'the winner takes all'. Ideally each dog should have his own kennel. There is truth in the saying 'dog eats dog'. I have listened to gruesome tales related by dog dealers about fights occurring when a number of young dogs were housed and fed together, unsupervised, where the end result was that the weakest dog was torn to pieces and eaten by the rest of the 'pack'. A dog which is housed on his own has plenty of time to think, and work up a keen appetite for learning and work. A dog which is bullied by another will never show his true potential either at home or on the trial field.

Treated with patience and kindness a puppy will grow into a confident adult.

It is often said by below-average sheepdog handlers, that good handlers have 'made' dogs. True, good handlers are better at covering up faults, but how many dogs exist without any? I believe that the more time you are prepared to spend with a dog, the better he will finish up. An extra six months or a year of patient training can be the difference between an indifferent work and trial dog and a great one. In the long run, it will be the genuinely intelligent dog which will make a name for his master, and not the other way around.

In play, puppies cultivate many of the balanced movements they require to work sheep efficiently when they become adults. They also benefit greatly from being exercised from an early age on rough hill ground to ensure that they become sure footed.

A few years ago we acquired a four-month-old puppy which had never been for a walk and had spent most of his time shut in a small shed. There was quite a difference between him and our home-bred six-week-old puppies when they were all outside. The newcomer had to be lifted over clumps of grass and large stones and I rescued him from the burn where he took a flying leap on his very first outing. At the time, I found this incident amusing, on reflection I find it rather sad. If puppies are allowed freedom from an early age it saves an awful lot of work later on. However, it is worth remembering that a large puppy is not quite so nifty on its feet as its smaller contemporary, and takes longer 'to pull itself together', as we say.

GWB Jr and friends. Partners in Grime.

The human voice plays a very important role in the early weeks of a puppy's life and he should be spoken to whenever possible. The tone of the voice is quickly associated with either praise or punishment. The latter never exceeding a mild shake or a severe scolding, which is reserved for cases of extreme naughtiness (eg, poultry worrying, chewing household fixtures and fittings and peeing on same!).

Puppies treated with patience and kindness grow into adulthood with an inbuilt confidence, whereas those which are rarely spoken to, handled or taken for walks, are often reluctant to leave their master's side when training commences because they lack the confidence to do so. Dog puppies are sometimes slower to mature than bitch puppies, however, they are often more dependable when they become adults because they are less temperamental. Mainly due to temperament, very few bitches have won an International.

Choosing and rearing puppies

Two puppies are as easy to rear as one! It is unnatural for a puppy to grow up alone; an only child finds companionship at school and, just like children, puppies need to play, quarrel and generally let off steam in order to develop into well balanced adults.

When purchasing puppies it is advisable to choose a dog and a bitch from different litters, preferably with the same powerful hill blood and, where possible, with at least three or four generations of it, further back in their pedigrees. Choose puppies off hill parents, (preferably dogs which can work without constantly having to be told what to do) as your foundation stock.

Buy your pups from a shepherd or farmer who has kept the same line going, for a number of years, obviously he has found them to be sound and capable. Determination and sensitivity are useful attributes in a dog. Dogs of this calibre are usually powerful, intelligent and reasonably easy to handle.

Be sure that the puppies are eligible for registration with the International Sheep Dog Society and their parents have been passed clear of Progressive Retinal atrophy or PRA.

Pups should always be chosen from nicely marked, well fed litters. Personally, I would not choose an over-bold and friendly pup. If living in the wild it would welcome a predator, and probably not survive for long! I prefer a pup which is calm but cautious.

Puppies should always have the insides of their hind legs inspected for dew claws as those that have them or have had them tend to be cow-hocked. Check the mouth to see if the puppy is undershot, the teeth ideally should meet, but many are fractionally short—if it is only a fraction I would not worry unduly.

Black lips and noses look far better than pink ones, and are said to be a sign of good breeding and hardiness, as is a black roof to the mouth. I have not been able to substantiate the latter statement. Plenty of space between the ears is often a sign of high intelligence. The top of the head should be long and flat in profile. A large bump of knowledge towards the back of the skull (not noticeable in a small puppy), is often found on a brainy dog. Dogs which have a pronounced stop, where the muzzle juts out from the face, are often more amenable than dogs with a badger-type head and little or no stop. A short tail is preferable, as too much can go wrong with an excessively long one. I do not care for a puppy with a wall eye (blue eye) as I am firmly convinced that their vision is impaired by bright sunlight. Often they will close the offending eye on a bright sunny day.

When possible, puppies should be purchased in the spring or early summer in order to get some sunshine on their backs and for them to grow bigger before the onset of winter.

When collecting puppies they feel more secure packed with straw in a dark, well ventilated box than rattling about in the back of a car. Enquire about their diet, as a puppy's stomach is easily upset. Ideally they will travel much better on an empty stomach and on arrival at your home should be given nothing but a drop of warm milk until the following day. Any feed that is different to the kind they are used to should be introduced very gradually.

The puppies' bed should be warm, dry and free from draughts, with plenty of light, and there should be no chinks or cracks in the kennel door. (Believe it or not, I have seen puppies who have developed squints from spending hours on end peeping through an aperture in their kennel door.) If puppies are given three balanced meals per day, clean water, plenty of milk, as many eggs as you can spare, regular exercise, rides in the car, fun and games with the children—you will reap the benefit later on.

Once puppies reach the three-month stage, they will require their Parvo virus

Above *Tweed, brother to Garry.* **Below** *Jan ('the Bionic Bitch') and GWB working together.*

Right *'Set to drive'. Laddie at ten weeks.*

innoculation followed two weeks later by a three-in-one vaccination against the dreaded Distemper, Leptospirosis and Viral Hepatitas. Personally I leave the three-in-one vaccination to a later date because I have seen some dogs become very much under the weather having had everything included all at one go.

After yearly booster injections, the adult dog may appear lethargic and under the weather for a day or two. He must not be considered 'shirking' his work but merely incapacitated! The same applies when he is casting his coat. It is not generally appreciated how much this natural occurrence affects a dog, both mentally and physically.

Nothing equals the heartbreak of losing a dog with which you have been able to create a wonderful partnership. The memory may dim but you never quite forget, there are too many reminders. During the past four years, we have lost two of our dogs under very sad circumstances. I will describe these so that perhaps in doing so other deaths may be averted.

Tweed was a black and white rough-coated brother of Garry. As a trial dog he was extremely promising, being placed eight times in 11 outings during his first 'season'. After working for many days in severe blizzard conditions, Tweed became dull and listless and so he was wrapped in a warm blanket and driven the 12 miles or so to the nearest vet, a prolonged and hazardous journey, taking many hours to accomplish due to the deep drifting snow.

The diagnosis was Leptospirosis of the kidneys which was probably contracted from his being in contact with the urine from the many diseased and starving foxes which were roaming the countryside and foraging for food in the farm buildings (owing to the extremely hard winter). Tweed's annual booster against the disease was a month overdue because of the weather. Despite all efforts to save him, Tweed died later that evening in his master's arms. It was two days before the storm abated but in spite of the arctic conditions, we buried him in the frost-bitten earth, wrapped up in 'his' blue blanket, near a sheltering stone wall where, hopefully, come spring, crocus and daffodil would bloom. As I walked back to the warmth of the kitchen, I wondered why it is always the best that have to die. I think of Tweed now, lying in the shadows, his bright eyes like pools of amber light, soul searching, watching the ewes on the hill, anticipating their every move.

Jan was one of the most competent trial bitches which GWB ever had in his possession. Her breeder was Les Morrison of Cleugh Brae, near Otterburn. Her dam was his bare-skinned Jean and her sire was Spot, a powerful rough-coated grandson of John Gilchrist's Spot, a Scottish champion. She came to the Bowmont Valley when she was three months old and with her sleek head and large eyes, she bore a strong resemblance to a young otter I had watched as a child from a leafy hideout on the banks of a slow flowing river.

Because of her tremendous speed and great skill in turning 'on a sixpence', I gave her the 'pet name' of 'Bionic Bitch'—a name she certainly lived up to! Running at a trial she was so beautiful to watch that she took your breath away as she skimmed effortlessly over the emerald green parkland, working all sorts of sheep with a calm precision that had to be seen to be believed. Spectators were often heard to comment that she bore a strong resemblance to a greyhound! Not only was she a joy to watch but, according to her master, she was also a joy to handle, never in her life questioning a command if it was justified. Jan won and was placed in trials all over the country, winning three trials in succession in one glorious weekend of competition in the borders.

In the late summer of 1980 Jan started to lose flesh and her abdomen appeared to be distended. Her normally silkey coat lost its lustre. After a couple of days her appearance returned to normal and we put her symptoms down to a false pregnancy. At Christmas her trouble reappeared and we immediately took her to the vet, who inserted a syringe into her abdomen. The news was bad—her abdomen was full of blood, due to internal bleeding.

Within the space of an hour our beautiful Jan lay dead. She died of a severe haemorrhage from the liver, during an operation to save her life. The cause of the haemorrhage remains unknown although we wonder if her speed and fast turns were a factor, or perhaps she received a knock from a cow on the hill, unknown to us. She lies across the way from Tweed in the shade of the mountain ash. A moss-covered boulder marks the spot which is inscribed very simply 'Jan'.

It is important that puppies are checked regularly to see that they are free from fleas, lice, mites and ticks, all of which cause discomfort to the pup and stop it from thriving. Both the kennels and the puppies should be regularly dusted with louse powder (with careful attention given to the base of their ears). Mites will take refuge in a dog's ears and have been known to cause infection.

Gentle scratching of the ears or shaking of the head always warrants a look, just in case the dog has canker. In the early stages there is often nothing visible and you will need to look again the following day. Sometimes it is merely earache. If the ear is discharging or inflamed, your vet will clean it and prescribe the appropriate cure.

Puppies can be very prone to stomach upsets, too, attacks of diarrhoea can be caused by worms, enteritis, a change of diet or a car journey. Never hesitate in seeking the advice of a vet. To delay allows the puppy to dehydrate which can lead to death.

Parvo Virus is another cause of diarrhoea, accompanied off and on by feverishness and a high temperature. In some young puppies this disease is fatal if not recognised in the early stages, and the correct amount of liquid and anti-biotics administered. The symptoms are a very slight discharge and a pink tinge to the whites of the eyes, followed by vomiting and severe, foul-smelling diarrhoea.

Telephone your vet if you have the slightest suspicion of Parvo. It can affect dogs of any age, although some show no symptoms. Others can recover within a few hours, or days. The very young and the very old, should they develop this dreadful disease, are often the worst affected. Parvo is extremely virulent and will travel on feet, clothes and car tyres. Pregnant bitches pass the disease on to their unborn puppies, some of which can die of heart attacks at around the age of 11 months. A severe scrubbing and dousing with a strong household bleach is one of the most effective precautionary measures.

Early days

And so . . . you have your pups! They have been named and are healthy, well fed, warmly housed, clean and, hopefully, raring to go! They will require exercising at least twice a day. Ideally, they should be allowed to accompany you and an older dog to the hill to learn what life is all about. If a puppy gets a drenching unavoidably in cold weather, he should be dried off with an old towel or piece of sacking. In stormy weather it is equally important to see that an older dog is rubbed down before putting him into his kennel with some dry straw. (Imagine if we were sent to bed in wet clothes in the winter!)

When out for a walk it is advantageous to call the puppies to you frequently, using their names. You must encourage them to come quickly, and be lavish with your praise when they comply. Introduce them to a 'come here' whistle of several short notes, to bring them to your foot. When you return the puppies to their kennel, ask them to 'stand' and give a stop whistle, one short whistled note, before closing the door, so that when the time comes to visit sheep on a regular basis, they will have a rough idea what the command to 'stop' means.

At around the age of three months, they should be introduced to a small, well behaved flock of sheep in a suitable well fenced area, where the following reactions are perfectly normal and acceptable—they split the flock and chase the sheep, grip the sheep, run away from the sheep, bark and then run away or they set in to drive, showing eye and style, or gather the sheep and bring them to you (a rarity).

It is imperative that the pups be allowed to 'do their own thing' as interference at this stage may cramp their style, particularly when a puppy is of an over sensitive nature. Great care is needed to ensure that they do not get hurt or become frightened when they first see sheep or they may never go near one again. After the second or third visit, once they have got over the initial excitement, you will begin to see something of their true characters emerge.

Puppies, naturally, are often reluctant to leave the sheep when it is time to go. It is therefore advisable after the initial visits to only take one pup at a time, as one is much easier to manage than two (just like children!) When they do refuse to leave sheep, try to arrange it so that the sheep run into a corner, and then walk away. Give your 'come here' whistle and call their names then leave the field and stand out of sight. Within a few minutes most pups will come haring into sight, tongues out, eyes bright and tails wagging. Praise them greatly and tell them: 'That'll do'.

Pups which flatly refuse to 'come away' and those that 'break' away to the hill or into the next field will have to be caught and either carried or put on a string and only released when you are well away from sheep. Personally, I admire the pup that is constantly breaking away to sheep, though I never let 'him' know it!

The majority of youngsters go through an annoying phase of not wanting to go into their kennels. Just like human children, they think that bedtime is a bore. If you walk on past the kennel door most puppies will come to you if you speak to them kindly. When they become wise to this ruse, entice them into the kitchen or slip a string on to the pup before you reach home, and at a different location each time. Never endeavour to chase a puppy in order to catch him as you will only frighten him and, anyway, he is much faster than you are. If this kind of naughtiness reappears when the pup is older, put him on a chain and insist that he goes in and comes out of the kennel several times with you giving the appropriate command on each occasion. The same lesson applies if he refuses to get inside the car. Always give plenty of praise when he does what you ask of him.

Never, under any circumstances, leave puppies, or any dog for that matter, tied to a fence. It only takes a very short time for any animal to strangle itself. Dogs can so easily become entangled with one another or around a tree or post. They may also leap over a wall or fence and be left hanging on the other side. I speak from experience. When I first began shepherding I owned a young dog called Fleet. He was a bonny black and white bare-skinned son of the tempestuous Shep (the dog which bit Ernie Crisp's ankle on his own doorstep!)

When not working, Fleet was kept on a long chain inside a large building. Behind him, high up in the wall, was a small window with a narrow ledge. The window had been firmly boarded up for a number of years to prevent draughts. It was therefore quite forgotten. Young Fleet was an intelligent, enterprising dog whom I cared for deeply. One day when I had set off with his sire, Shep, to gather some sheep, Fleet, disappointed at being left at home, somehow managed to leap up on the ledge where he removed one of the boards and, in trying to jump down the other side, hung himself. Poor Fleet died, not because of my lack of care, but because of my lack of thought. I have never chained a dog up since that awful day, and never will again.

Basic training for puppies

Serious training should not be attempted at too early an age. Training when the pup is between eight and 12 months is quite soon enough and the final polish can be added at around two and a half years of age. Generally the older the dog, the more proficient he will be with sheep.

Basic training in the early months of a puppy's life will benefit it later on. When out walking with the pups ranging ahead, you will notice that often when you stand still, they stop and look to see what you are about to do next. They are naturally inquisitive and so give you a golden opportunity to tell them what the word 'stand' means. Quickly, before they have a chance to move off, ask them to 'stand', then give them the 'come here' whistle. Encourage them to come quickly and praise them when they reach your foot. Repeat the lesson at regular intervals.

Pups tend to learn very quickly and are generally more intelligent than we give them credit for. Once they grasp that you wish them to 'stand' at a distance from you, it should be relatively simple to get them to stop dead in their tracks when they are running towards you. If they keep on coming, go towards them with your hand raised, repeating the command. Encouraging the pup to come to you at speed will be useful when he is older and you want him in 'post haste' to stop a galloping ewe when shedding.

The 'come here' whistle has many uses; for instance it can be used on the cross-drive to encourage the young dog to come quickly in your direction, before he is sure of his flank whistles. You can also use it to pull a dog which is inclined to flank tightly, off his corners on the down drive. On the fetch, when sheep are coming helter-skelter, it can be used to encourage a wide flanking dog to go tightly down the sides of the sheep, and in front of them, should it be the only way to halt their rapid progress.

As long as a young dog will stop when asked, I never insist that he lies down, much preferring him to stay on his feet. There will come a time when, because of the sheep's behaviour, he will be required to lie down in order to pacify them, but in such situations 'eye' will generally naturally make a dog go down. A dog which works on his feet around a trial course not only looks smoother but also keeps his sheep quieter and more settled than the dog which is constantly stopping and starting, bobbing up and down like a demented frog. I can see nothing to justify teaching a young dog to lie down in the early stages of training, and then having the tedious task of putting him on to his feet again later on.

It is foolish to expect any young dog to behave consistently all the time, as all of us have our 'off' days and dogs are no exception. A dog's 'off' days are best spent in the solitude of his kennel. After a day or two's rest you will be pleasantly surprised at how refreshed he will be when next you take him to sheep.

Sons of Garry becoming familiar with sheep.

Above *Dog and lamb—study in harmony.* **Right** *Dogs and puppies cannot spend too much time in your company. GWB and his dogs.*

Once a week, or even less, is enough for a sensible three- to six-month-old pup to visit sheep. A delinquent 'gripper' needs to be kept away until old enough to tolerate a reprimand. The average well brought up puppy, if given the opportunity to do so, will outgrow faults or will eventually allow them to be cured by a trainer he has learned to trust and respect.

When the dog reaches training age, usually from eight months onwards, it is vital that he enjoys his lessons which must not be prolonged to the extent that he grows tired and bored by them. Working him in a different field each time you take him to sheep will help stimulate his interest or, better still, take a few sheep 'walkabout', saying little or nothing to the dog and thus encouraging him to develop initiative and confidence in his own ability. If a young dog does become completely disinterested it is advisable to keep him right away from sheep for a few days. When you do return the company of a trained dog will often ensure that his keenness returns.

You must make allowances for a dog's youth and inexperience. Rome was not built in a day! Any serious faults which occur must be corrected gradually, without the dog realising what you are doing in case you upset him. You must always try to remember during training sessions, especially when nothing is going to plan, that what is easy to achieve is often of little value.

If, by chance, you and your dog do have a misunderstanding which leads to

him being taken home and put in his kennel, do remember to make your peace with him before bedtime. Facing a long lonely night must be a daunting prospect to a young dog in disgrace. If you are kind, but firm, on all occasions, a bond of trust will develop which will only be broken by you being constantly unreasonable during training sessions. A puppy cannot spend too much time in your company or the company of your family. Excursions to a well behaved flock, lunch-times spent in the kitchen, car rides all help to cement a good relationship.

On sighting cattle, pups will often go haring after them, barking their heads off. Do be on the watch as it is so easy for a limb to be broken or an eye to be put out by the kick of an irate suckler cow. Sometimes contact with cows and calves is unavoidable. When it happens, try speaking quietly and kindly to these fractious beasts, shouting only convinces the dog that you are encouraging him to attack them.

It is the height of folly to use a sheepdog for the purpose of working suckler cows. I have seen quality dogs maimed and killed doing just that. Only certain dogs, often the rough, commoner types, acquire the specialised skills required to deal with an angry cow with calf at foot. If, for safety's sake, a man needs to take his dog among calving cows to protect him, the dog should be told to keep well in the background out of harm's way and should be called only if absolutely necessary.

Early last summer, on a gloriously sunny day, GWB and little Trim set off to a poor part of the hill to bring home a black cow and her calf. Despite the fact that Trim flitted about like a shadow, at least a hundred yards to the rear, the cow waited until GWB turned his back to open the hurdle into the haugh before she made her move. She knocked GWB to the ground (which, fortunately for him, was wet and yielding) and trampled him underfoot until he lost conciousness. When he came round (luckily no bones were broken, although he was badly cut and bruised) Trim was lying by his side licking his hand with her warm tongue. The cow and her calf had gone from sight, we assume that in some way Trim had distracted her but we have no proof other than the fact that she was muddy and dishevelled in appearance and exhausted at the time. This ordeal showed GWB how terrifying these experiences must be for a dog.

Quiet handling is nearly always more effective with the trained dog as he will need to give you his attention in order to hear your commands. It is imperative that you always speak quietly to your young dog, so that when it is necessary to scold him it will come as a shock and be the only punishment he requires. Some people rant and rave continuously until their poor dogs don't know whether they are coming or going.

A strong, impetuous dog especially benefits from quiet handling and should a dog develop the unpleasant habit of gripping sheep (an inherent trait), it will be necessary to endeavour to cure him with a non-drastic method. Stand in the corner of a field, encouraging your dog to work three sheep tightly up to you. He is now in a position where, should the sheep break, the excitement will make him forget his manners and he will attempt to grip. You must be ready for him, and chase him back up the field scolding him severely with your tongue! Encourage him straight back up to the sheep immediately. Should he again attempt to grip (you will see by the expression in his eyes when it is about to happen), quietly say to him: 'Now don't you bite them'. Eventually he will come to realise that the only time he is allowed to take hold of a sheep is if and when you say so.

My Garry never attempted to take hold of a sheep until he was 3 years old. He was a determined dog who could always move his sheep. However at lambing time, if there was a ewe which needed assistance, Garry simply refused to catch and hold her for me, a great inconvenience. One afternoon, while working in the sheep pens, a particularly fierce old ewe cornered him and, before I could intervene, butted him hard against the 'keb house' wall. Without the slightest hesitation, he grabbed her by the neck-wool, tipping her cleanly upside down. I praised him, and since that day he has become an expert in the art of catching a single sheep and holding it.

Just like humans, dogs vary enormously in their make-up. That difference must always be taken into consideration and allowances be made for it. Sensitive, easily-handled dogs must be given a free rein in order to build up their confidence. Training with this type must advance more slowly, preferably when the dog is older and wiser. Training at too young an age will make a 'soft' dog less inclined to use his own initiative.

The less initiative a dog shows, the longer you must allow him to do as he pleases. Ideally he should just be taken to work and allowed to run alongside an older experienced dog. In the case of the easy dog let 'least said, soonest mended' be the rule. To a certain extent this rule also applies to the determined dog. Too rigid discipline with this type of dog will make him run amok at the sheepdog trial. It is very important that he be allowed to gain confidence in his own ability. To do so he must be allowed, whenever possible, to learn by his own mistakes. Plenty of hard work is the best cure for his wilfulness.

Young dogs which gather sheep 'straight', chase and split the flock, often end up with the most power and determination. Pretty workers which lie back off sheep, giving them plenty of room on the corners, look marvellous but they can grow up to be weak and ineffectual. Pups which work on to their sheep are to be preferred. I would much rather have to keep stopping a dog than ask him repeatedly to 'walk up'.

Occasionally, you will have dogs which do not make the grade, either because of their natures or because they will not work at all. These dogs should be treated kindly, not with contempt, and you should do everything you can to find them an agreeable home.

A bond must form between dog and master (or mistress!)

Trained dogs, surplus to requirements, will have to be found suitable owners, and they also deserve consideration. It is always a sad occasion to see dogs sold. I console myself with the thought that they have been treated well and, just like children, that they must eventually leave the nest.

Although some dogs will only work for one person, others will quite happily work for anyone. This sort often benefits from the change of ownership. When a sensitive dog goes to a new home, he will often lose some of his confidence which only time and understanding can replace. Because of fear and insecurity a sensitive dog will often perform the most foolish acts, completely out of character, and therefore quite unexpected.

Fear manifests itself in numerous ways. Dogs which are normally particular about their appearance suddenly start to roll in all kinds of filth as though trying to hide their identity, owing to a feeling of insecurity. They foul their kennels and, when asked by their new owner to gather sheep, may leap over a fence or wall, refusing to work. Eventually, after a suitable period of time has elapsed and a bond has been created with the new owner, the dog returns to his normal self, sometimes becoming a better dog than he was before.

A dog which regularly changes hands cannot be blamed for not giving his all and for not allowing a bond to develop between him and his master. He will be in a permanent state of insecurity. Anyone purchasing a dog should be told absolutely everything about its character so as to make the transition as least traumatic as possible.

Some useful advice

Speaking from much regretted experience, it is the height of folly to leave a young dog alone in your car. When you return there is every likelihood that the upholstery will be ripped to shreds, the collie being an expert in the art of demolition. For his first few outings, it is always advisable to take an older dog along, or leave someone in the car to keep him company. It is always a good idea, when leaving dogs in a car, to park with the rear window facing your direction of retreat. Hopefully, the dogs will sit quietly and watch for you to return.

It is pointless punishing a dog after an event—the same rule applies for any misdemeanour. Unless you actually catch him in the act, he will have no idea what he has done wrong. It is much better to muzzle a dog until he loses his fear of being left alone (or, in the case of a 'real' dog, his anger at being left alone!)

When the weather is hot a dog should never be left in a car for long periods. Where possible in warm weather, park in the shade and leave windows fractionally open. Water should always be available and the dog allowed out frequently for exercise and to relieve himself.

You cannot better raw meat for getting a working dog keen and on his mettle (provided he is regularly treated for tape worm). Meat to the dog is what oats are to the horse! (The shepherds of old wormed their dogs by giving them rotten mutton with the wool still on it! I cannot vouch for its effectiveness, and would worry about the occurrence of a blockage, so stick to pills!) As long as a carcase is free of drugs or poison, it can be skinned and suspended above the ground so that the dog, when peckish, can help himself. At lambing time it only takes a few minutes to skin a stillborn lamb. It is at busy times like these that the dog will benefit most from natural food, but do include a little of his normal balanced ration.

Teething
Puppies, just like human babies, have teething troubles, resulting in irritability owing to sore gums. A large solid bone to chew on will help distract their attention away from the kennel door! As a result of teething, a puppy may throw a fit—excitement can bring about convulsions. The puppy may be normal one moment and the next be lying, kicking on its side and frothing at the mouth. After a while he will rise and look around in a dazed manner. If not caught he may gallop off, often barking, quite unaware of his surroundings. It is advisable to put him in a quiet, darkened place and consult your vet.

Annoying habits
If you have a bitch in season, it is a natural reaction for any dog around to behave disruptively. His work will suffer and therefore you will have to be prepared to make allowances.

Some pups and young dogs develop the annoying habit of hanging on to the neck of an older dog. As long as the dog does not object, I do not either. Like most habits, this one will die out in due course, if ignored.

Around the age of 18 months any young dog worth his salt will endeavour to 'beat you at your own game'! The dog at this stage, in his mind, is reaching maturity and wants to be boss. All the best dogs go through this adolescent phase. Your dog must not, under any circumstances, find you lacking.

It is usually at this time that you may be tempted to part company and sell him. If you resist the temptation and give him another chance you will rarely regret your decision.

Eventually all that you are aiming for will slip easily into place. Most dogs improve in leaps and bounds at around the 2 to 2½-year-old stage. There is one exception to the rule, and that is the 'grass chewer'. This dog, because of temper, nerves (or both) pulls mouthfuls of grass and spits it out again all the time he is working. This obnoxious habit is hereditary and if the owner is wise he will not breed from this dog.

The rope trick
The disobedience I refer to in the 18-month-old dog usually manifests itself in two ways. All of a sudden, the dog will neither come when called nor stop when or where you tell him unless he feels inclined to do so. You will have to match his determination and disobedience with your cunning and come out 'top dog' without losing your cool.

The moment the disobedience occurs, take him home and attach him to a long cord. Firmly ask him to 'stand', and then walk away a short distance. Call him to you in a kind tone. If he refuses to come pull on the cord. Repeat the lesson as often as it is necessary. Eventually, when he comes to you willingly, praise him immensely.

When the dog refuses to 'stand' tie him on a long cord to a post. If he makes to follow you, keep taking him back to the post and repeat the command firmly. If this does not work, tie him up short, ask him again to 'stand' and walk away. Walk round about him for a few minutes before extending the cord. When he does as you ask, allow him to trail the cord. Ask him frequently to come to you and to stand. Enforce the commands if necessary. It is highly unlikely that you will need to, because he will by then have realised that you mean what you say, especially if you are lavish with your praise when he complies.

The following day you should be pleasantly surprised by a dog who has fully accepted you as his pack leader. This kind of problem usually only arises with a determined type of dog who, in the long run, will prove the most dependable and wear longer than a soft dog.

Chapter 11

Sheepdog training

The trainers—'makers or breakers'?

There are those dog trainers who 'make' dogs and those, sadly in the majority, who 'break' dogs in a military fashion, never giving a thought to their feelings, either because they are impatient for results, or because it is the only way they know, owing to a lack of imagination.

Sensitive and artistic people recognise that there is an art in training and handling a sheepdog. Because they wish to 'make' or create something beautiful they quite often get 'carried away' (sometimes amusing the spectators by quite unintentionally 'putting on the style'). They are alone in some far off magic land with only the hills, the sheep and a dog who is creating 'poetry in motion' before their very eyes. Unfortunately these people are few and far between.

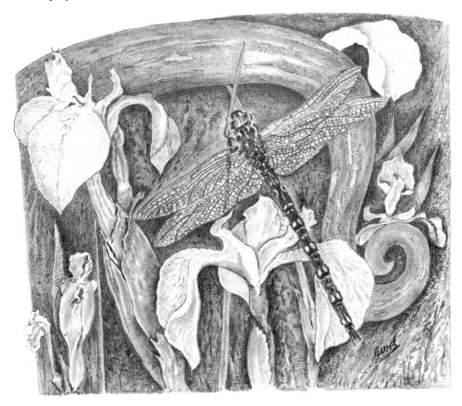

The category under which one comes, I think, depends largely on one's make-up and the way one was brought up. 'Making' a dog is a lengthy process and can take up to five years or, as a much respected Scots handler put it: 'A trial dog is at his best when he has a year's experience under each leg'.

Trainers must be prepared to create circumstances regularly where a dog can do what is required of him, in a natural happy fashion. More effort is required on the part of the trainer but the end result makes it all worthwhile. Instead of being the owner of a slave, you will be the friend of a capable, smooth-operating dog with his dignity intact.

'Breaking', put in a nutshell, is relatively quick and straightforward because the trainer simply insists that the dog does everything he is told, when he is told and eventually how he is told, whereas 'making' a dog entails adapting to the dog's ways and fitting in your commands with whatever he is doing whenever possible!

The plain worker

Owing to today's average trialling conditions which consist of wild sheep, run on a medium-sized course, a plain working dog is favoured by many sheepdog handlers. Obviously the less 'eye' a dog has the easier he will be to flank about. Plainness in itself is acceptable if partnered with brains and power. A dog with a plain method of working, well handled, can look classy and intelligent, badly handled, he is very boring to watch. The less a dog sees of sheep, and the fewer sheep he sees, the more 'eye' he will gather.

Below *Two dogs operating smoothly. Lucy and Laddie.* **Right** *Lucy using her own initiative!*

A plain dog usually alters with age—by the time he comes to his best, at around 5 years old, he will most probably have developed the required amount of 'eye' to look good. A plain worker which is lacking in power is not a good proposition. Unfortunately there are a number of dogs like this as they are bred for their soft natures and easy handling with regular trial winners of a similar type. The plain dog which is lacking in power, can win and be placed in small trials, however, under rigorous conditions, on a large course in bad weather or with awkward sheep, both dog and handler will have a difficult time beating the clock!

Working an 'easy' dog gives you plenty of opportunity to eavesdrop on the critics. They will not show you any sympathy! It is often necessary to 'over command' an easy dog to instil confidence and keep him moving. However, when working at home, he will benefit from being left to get on with it so that he can develop his initiative.

It is pot luck as to how much 'eye' a dog will develop as he matures. A dog with a very strong 'eye' which lies watching sheep as though mesmerised by them, will benefit from a nudge with your toe if he repeatedly ignores your command to 'walk up'. A dog with too much 'eye' will improve if he is worked regularly in the sheep pens, and on large flocks. A shortage of work will only ensure that he gathers more 'eye' and becomes stickier to move.

A particularly powerful and determined dog is best kept away from sheep for two or three days before a trial. The 'eye' he will gather due to his enforced rest should slow him up sufficiently for you to get a controlled 'lift' which should help to keep him settled and, hopefully, under control throughout the run.

The advantage of running a powerful dog is that you are so busy keeping on top of him that you tend to forget any nervousness you may feel, almost believing you are on the 'green grass of home', with only the sheep and birds as spectators.

There can be a fairly common and quite dangerous problem with a keen excitable young dog, occurring mainly in hot or humid conditions, which takes the form of staggering about and sometimes falling to the ground, usually after work or a training session (and not due to alcohol!) This condition is less likely to happen to a dog which is fit and of a calm disposition.

Because a dog perspires through his tongue he cannot always rid himself quickly enough of the toxic waste products in his body. It is vital to get the dog to a cool burn or a water trough at the first sign of trouble. If you do not it is possible for him to lose consciousness and die. You must not allow him to drink too much cold water at once, merely cool off in the water and, should this problem persist, you should ask your vet to check the dog over.

Feeding on a quantity of raw meat can affect a dog's temperament, in that he becomes excitable and bad tempered due to the high protein intake. This does not occur as long as he gets plenty of work to do. Dogs thrive well on tripe (if you can stand the 'healthy aroma'!) However, great care must be taken in cutting them into small pieces to ensure a greedy dog does not choke.

Patience and perseverance

Patience and perseverance is the golden rule when a young dog's training commences. It is more interesting for a dog to be taught a little of everything that is going to be required of him, than to be taught one thing at a time. An absolutely genuine dog is difficult to spoil, but such dogs are few and far between. I prefer a dog which attends to 'his business' after work, and stretches when he comes out of his kennel. He is usually intelligent and relaxed. Working a dog in different locations, sometimes on a large flock, or on half a dozen sheep, will help to stimulate his interest.

Exceptionally 'wide' running dogs benefit from regular work within the perimeter of a field until they are at least 3 years old, only being used occasionally on the hill for short gathers and close at hand work. Powerful dogs with natural line greatly benefit from plenty of hill work. Wild, fast sheep running over uneven and steep ground teach young dogs to flank freely and keep their correct distance. If they do not, they will end up in the middle of the flock, losing most of the sheep in the ensuing confusion. Often performing difficult tasks with a dog makes a simple task, in favourable conditions, relatively easy to accomplish well.

The average youngster is ready to begin his training in earnest at around 10 months of age. Generally, the older he is the quicker he will learn what you expect of him when handling sheep. On average, bitches mature earlier than dogs, but due to their temperaments they require training with more consideration and patience. Every other day is quite sufficient to take a young dog for a training session, as the day between will give him time to collect his thoughts and allow what he has learned to sink home. If there is an occasion when he loses interest in his lessons, it is far better to take him home quietly and put him in his kennel where he will come out more inspired the following day, or even the next!

An unenthusiastic youngster benefits from visits to wild flying sheep, which should stimulate his natural instinct to hunt and pursue. Occasionally a dog will not even give sheep a thought until he is at least 2 years old or over. Thankfully these dogs are in the minority, and strangely, due to a magical transformation which can literally take place overnight, they often become 'the best dog' their master has owned.

At the age of 10 months a young dog should respect and know his owner fairly well. When called or whistled he should come quickly, he should have a good idea what 'stand' means and should leave sheep willingly when asked to do so. The 'one in a million' ideal dog has a willing nature, a natural pear-shaped outrun, a natural balance, will flank freely off either hand, will not overflank, will walk right up to a sheep's face without any sign of fear, is good tempered and not inclined to grip unless greatly provoked and works with a tight, perfectly set-on tail. If one dog of this sort comes your way in your lifetime, count yourself extremely lucky and do everything in your power to be worthy of such a great and wonderful gift.

The average dog either lacks power or has too much. The ideal is a quiet power which is, unfortunately, a rare quality. It is important not to confuse weakness which is a genuine fear of sheep with too much 'eye'. In time, with plenty of work and encouragement, 'too much eye' will leave a dog.

A very weak dog is afraid to look sheep in the face and is ready to move them with his teeth whenever the opportunity arises. Weakness comes in varying degrees. Allowing a weak dog to grip sheep will boost his confidence, but it is a cruel method to resort to and painful to the poor sheep who have to bear the brunt of these often unwarranted attacks.

Should a dog with power, presence, determination or whatever you care to call it appear on the scene, sheep automatically move away from him. Power is immediately obvious, but like weakness, there are varying degrees of it in different dogs. Quiet power is recognised as an authority over sheep which does not frighten or upset them.

Whilst much, in time, can be achieved with a powerful dog, very little will be achieved with a weak dog. The latter deserves to be treated sympathetically. Help him to move awkward sheep, chide him gently when he grips and, who knows, most dogs acquire confidence with age. He could make a useful work dog. Because a chain is only as strong as its weakest link, this type should not be used to breed from.

Choice of commands

Before training commences you will need to decide on a different set of commands (both words and whistles) for each young dog. The equivalent whistled command must be as different as possible and also different for each dog, so that working together they will not become confused or collide! Once chosen, the command must be adhered to rigidly. Eventually, when you have your dog fully trained, it will be possible to hold a conversation with him using whistled commands which vary in tone, speed and length, when he is at a distance either in fields or out on the hill.

When working a dog close at hand I much prefer to use quiet spoken commands. For his or her right flank, I say 'way'. For his or her left, I use the word 'come'. My second dog's commands are as follows: 'out' for the right-hand flank and 'bye' for the left. Both dogs are taught identical 'stop' and 'look back for more sheep' commands and whistles. Both are asked to 'walk up' but each dog is given a different whistle command for this.

I leave everyone to work out their own whistles for getting their dogs to flank, stop and walk up. On no account will I attempt to draw whistles! Anyway, my memory will only stretch as far as two different sets and that is on a good day!

At Swindon in the early spring, much to my amusement, the sweet shrill notes from the blackbird and thrush, courting in the dense nearby forest, often cause my dogs to leave my foot unexpectedly and veer first one way and then the other! (I swear those birds are great mimics and show off to their brides to be!)

Running brace

There is nothing to equal the beauty of two collies working in close harmony either within the confines of a trial field or on some distant bracken-strewn hillside. They and their charges appear to flow effortlessly along as though on the incoming tide, one dog counteracting the gentle pressures of the other, until at last the tumbling cascading mass arrives and with eyes questioning and nostrils quivering the sheep stand at your feet.

Above *Running brace. Two heads are better than one.* **Below** *Two dogs working in harmony.*

At lambing time it is a relatively simple task to acquire the habit of using two sets of commands. (A couple of years ago, after numerous lambings, I finally managed to master the habit!) All the established handlers of single dogs warned me that running brace was the quickest and surest way to ruin a dog. Naturally I wanted to know why and was given several equally valid reasons, the first being that instead of allowing the dogs to reach the 12 o'clock position when sent to gather sheep, it was necessary to stop each one fractionally short and in doing so, develop the 'points deductible' habit of 'stopping short' when running singly. The brace handler's defence was that the majority of them allowed their dogs to 'cross' once they got behind the sheep.

Other valid reasons against running brace varied from making the dogs 'sour' because it was often necessary to over-command them, to removing all their natural initiative so that they became mechanical in their method of working.

I did, and still do, worry about the effect that running brace has on my dogs. Unfortunately my enthusiasm remains. I was pleasantly surprised to discover that my 'duo' are for most of the time contented to stay on the sides I allow 'them' to choose! The impression I had previously was that the handler had a difficult task keeping the dogs to 'their' sides. I wonder whether closely-related dogs are telepathic and more in tune to each other, as GWB's litter sisters, Jed and Trim, have not been inclined to cross either.

An added surprise was that my most determined dog became much more amenable to handle. I can only assume the reason for this being that the other dog pushing the sheep on to him has the effect of slowing him down.

At home when I send two dogs for sheep, because I am genuinely worried in case they should stop short, I nearly always send them off the same side. Due to their slight jealousy of each other, they race flat out for the 12 o'clock position. Apart from the gatherings and at lambing time, I do not practise running double except on very rare occasions. The dogs, when at all practical and possible, are allowed to work on the sides they choose themselves. By allowing them that choice, I hope that they will neither become lacking in initiative nor have a preference for either the left or right flank.

I admit to practising shedding with them, pretending to 'pen' in a corner of the field. I found that using the dogs' names before, instead of after, a command was a good idea as my dogs memorised each others set of whistles very quickly. For us it is very early days, we have such a lot to learn in order to become a proficient team.

The outrun and lift

A young dog's training should commence in a small, well fenced area, preferably containing a score of well behaved ewes. Before reaching the sheep I usually ask the young dog to 'stand' and call him to my side a couple of times in order to remind him of what may be required when we reach our destination. If you own a trained dog, he will prove a useful asset at this time to prevent sheep from bolting into corners. Most sensible young dogs are capable of fetching them out on their own. If they do not know how to you will have to show them until they learn.

On entering the field, the young dog should see the sheep and run off in their direction. It is important that he is allowed to go without any interference from you. (Bite your tongue if necessary!) If he fails to spot the sheep, walk towards them and keep asking him: 'Can you see them?'.

Once he finds the sheep, the ensuing minutes will be spent getting the feel of things, running off excess excitement and generally having fun. If the dog plucks wool or grips the sheep you will need to intervene, but do not be too severe in your reprimand at this stage. He may, on the other hand, gather the sheep expertly and bring them to you without any training at all. However, there is every likelihood that he will either chase the sheep and split them so that they run in all directions, bark, do absolutely nothing at all or make to run home! All of this behaviour is acceptable.

Every young dog's reaction to sheep is different, unless he is ready to train you cannot start. It must be taken into consideration that some dogs mature at a much earlier age than others. The dog which runs down the middle of the field in order to retrieve his sheep is a better proposition than the one which hugs the dyke back. It is often much easier to 'put a dog out' than to fetch him in if he is inclined to run wide.

Once a young dog does settle down to work, he will learn most from being allowed a free rein. If his outrun is inclined to be straight, do not interfere for a few weeks, or until the habit of bringing sheep to you is firmly established. It does not matter how he gets them as long as he does! The outrun, lift and fetch at this stage should be accomplished by the dog only, with an occasional command thrown in, but not necessarily obeyed.

The dog must always be allowed to pick up sheep at the point of balance. Usually, but not always, this point is at the sheep's heads. Only the dog knows the exact point of balance. If the dog stops slightly short of the sheep, and if the sheep come away in a straight line, he must have been in the correct position to lift them and points should not be deducted at a trial by a competent judge.

Most young dogs, when they have run off their initial excitement, will look to you for guidance. The dog, once he begins to tire, will either stand or lie down, eyeing the sheep. This creates a good opportunity for you to fit in your stop whistle with what he is doing.

Once the habit of bringing the sheep to you is firmly established you have climbed the first rung of the ladder and can start to lengthen the distance that you send him. If you try to open him out on his gather, as he is leaving your foot during his early training, there is every chance that faults will develop, such as coming in tight or flat on to his sheep, or stopping short. To prevent the latter you will need to keep him on the move by either shushing him or telling him to 'keep back' until he reaches 12 o'clock.

If the dog does develop a fault, do not try to cure it overnight. If you are patient it may disappear on its own, otherwise you will need to create situations where it is simple for the dog to perform the task correctly. Later on, regularly keeping a young dog which is inclined to work tight, back off the sheep close at hand, will ensure that he will give them room when working at a distance. Some young dogs never make any attempt at all to run to the other side of the sheep, but are quite happy to drive any distance. You will have to decide on a way of getting him to 'kep' them (go round).

It sometimes helps to allow the sheep to bolt through a gate, or to allow him to run with a trained dog for a while. If all else fails, ask him to 'stand', and walk to the other side of the sheep. If you walk backwards, encouraging the dog to walk up, the sheep should follow you. If the dog makes to slip round the flock in your direction, stop him and move away round the other side. He will soon realise that his place is on the opposite side of the sheep.

To teach him to 'walk up' to sheep, place some in a corner then both you and the dog approach them from a distance. If the dog is in too much of a hurry, ask him to 'stand', followed firmly by 'take time'. Repeat, allowing him time to think, until he gets the message. Occasionally give him your walk up 'whistle' and be lavish with praise, as he progresses with his lessons.

Dogs with power and a natural ability to 'line' sheep up often have to be taught to gather correctly, and to keep back off sheep when they come in to the 'lifting position'. A free flanking dog, not over endowed with 'eye', often has a natural outrun and is usually more proficient at close-at-hand work, eg, penning and shedding. A dog which works on to his sheep is more practical for hill work or on a large trial course. You may not get your perfect 'pear-shaped' outrun but you will be able to put the dog exactly where you want him, and leave him to get on with the job in hand.

When weather conditions make your commands inaudible at a trial, the powerful dog, providing he has been allowed at home, will be able to line the sheep up on his own. Eventually the dog which gathers 'straight' will have to be 'opened out'. Short gathers, so that you can run between the dog and the sheep, are beneficial. You will need to chase him back off the sheep regularly until he gets the message and automatically opens out. Tell him repeatedly to 'keep back' until he does so on his own. Increase the distance you send him for sheep gradually.

Some dogs, from a very early age, will go great distances for sheep with little or no encouragement, whilst others show a reluctance to leave your foot! If, once you have set the dog off to gather and he makes to come in, stop him, allow thinking time, then, by using his name, attract his attention. Ask him in a gruff tone what he thinks he is doing, and with your stick point the direction you wish him to go. You may need to take a few steps in the appropriate direction, to ensure the dog understands your meaning and opens out. If he continues to 'come in', call him back to your foot, have a quiet word with him (it works wonders!) and set him off again. With some dogs, once or twice is enough to get them gathering correctly. With others it can take weeks or even months.

The ideal situation is a large flock which is spread out. Gathered regularly, it helps the dog to get into the habit of gathering correctly, naturally. Unfortunately for us this is not practicable with a hill flock during the winter months when they are heavy in lamb. At a later date, when the young dog knows his flank whistles, you can teach him that a short sharp whistle means that you require a fast tight flank by shushing him on. A long slow whistle informs the dog that either you require him to recast wider for another 'cut' of sheep or, if given fractionally faster, you want him to flank wide and slow on the sheep he already has.

An older, fully experienced dog, will eventually 'open out' after hearing only half of a slow whistle. When you start to teach the young dog to 'open out' on a slow whistle you must first of all stop him on his 'outrun'. Allow him thinking time, and follow up with a slow flank whistle. Keep repeating the exercise until he gets the message. If he fails to grasp what you require of him, walk up the field (or hill) and stand between the sheep and the dog. Walk towards him, giving your slow whistle. Keep calm, allow the dog to weigh up the situation and he will soon grasp the meaning of your slow whistle. If he is ready for training in the summer months, working regularly on the open hill soon encourages a dog to open out and give sheep plenty of room. Unfortunately for the dog, not everyone can provide him with this type of environment, so training can take much longer.

Once you have the dog casting out nicely from your foot, encourage him to get to the 12 o'clock position as quickly as possible. Numerous short gathers with you 'shushing' and telling him to 'keep back', will be beneficial in getting him quickly to the 'lifting' position. Sheep must be gathered regularly off both hands or you will soon have a dog which will only gather one way. The same rule applies at sheep-dog trials. It is tempting to send the dog the way he prefers and is best at, but inadvisable on every occasion.

A calm natured dog should be allowed to 'lift' sheep without stopping him. It is much nicer to see. At first, excitable dogs benefit from being asked to stop, and after a few seconds have elapsed, being allowed to 'lift' the sheep with a 'take time' command. A dog which 'hits' the corners of a flock, because he flanks 'straight' must be taught to 'bend off', otherwise wild 'trial sheep' will be further upset to the extent that they take fright, galloping first one way and then the other! By the time you come to 'pen' it will be virtually impossible to convince them to go inside! To cure the dog's 'straightness', encourage him to hold half a dozen quiet sheep up to you. Flank him first to the left and then to the right, bearing in mind that some dogs are only 'tight' on one flank. When the dog runs to 'hit' a corner, lean over the sheep, and flick him away with the end of your stick (this is easy to accomplish without touching the dog!) at the same moment quietly tell him to 'Gettt' with the accent on the 'T'.

You may need to run through the flock a few times in order to ensure that the determined dog turns his head off the corners. Because you have done your groundwork correctly and have a trusting relationship with the dog, this lesson should not offend him at all, and it should not be long before he gives the sheep 'room' on the corners.

The fetch

The dog must regularly be allowed to fetch the sheep without commands. This will allow him to find his 'distance'. By 'distance' we mean the position the dog must be in to handle the sheep authoritatively without upsetting them. The exception is the dog which weaves backwards and forwards behind a flock; he must eventually be stopped and kept well back, until he learns to follow straight.

A dog with excessive 'eye' and 'line' can become so engrossed with the sheep 'on the end of his nose' that he disregards the sheep on each side of the flock. I have often seen young dogs walk right through the centre of a flock with the two

or three sheep which are directly in front of them. When he does this he must be asked quietly what he thinks he is up to and put back for the rest of the sheep.

Some dogs are born with a natural 'distance', others find theirs. The remainder will continue to chase sheep until they are taught the most important command of all, which is 'take time'. To teach him this he must first be allowed to chase sheep and then asked to stand, followed up by 'take time' in a firm voice. If he moves off after the sheep at speed, the lesson must be repeated, perseverance being the key word! It is unfair to expect a young dog to go slowly continually. He must, on occasions, be allowed to run riot!

It is on such an occasion that you can use the opportunity to take him by surprise. Run quietly up the field, through the flock, and chase him back off them, at the same time asking him what he thinks he is doing! When you suddenly appear out of the blue, you will find it difficult not be amused by the surprised look on the dog's face, you can find it difficult not to laugh! (I always do, with the result that the dog thinks it's all a game.)

Throughout training the tone of your voice is very important. It shows the dog the difference between right and wrong, and it is by far better for the dog if the trainer goes to the extremes with his voice rather than his temper!

Teaching a dog to stand and to work on his feet

A dog which continually lies down when working is not as desirable as a dog which keeps on its feet. The 'upstanding' style upsets sheep less, and is more pleasing to the human eye. Some dogs naturally are inclined to work on their feet, whilst others have to be taught. From an early age the former, when asked to 'stop', tend to do so in a sitting position, gradually developing the habit of 'stopping' and working on all fours! When serious training begins, if the dog will 'stop' when asked there is no necessity to make him lie down. He should always be allowed to choose his own position. The dog which is inclined to lie down will require teaching to stay on his feet.

At a trial, there are only two occasions when it is necessary for a dog to lie down—they are at the pen and prior to the 'shed', and only then if the sheep are wild and completely unmanageable. To encourage bold sheep into a pen, the dog is best on his feet, looking them squarely in the eye. He should be capable of walking them in backwards if required to do so!

Provided a dog is not over endowed with 'eye', it should be a relatively simple task to teach him to work on his feet. The more 'eye' he possesses, the more inclined he will be to 'flatten' to the ground, and the more time and patience it will take to get him out of the habit. A relaxing and rewarding way of teaching a young dog to 'stand' and at the same time creating a loving bond, is to spend any spare moments grooming him. At first embarrassed by this procedure, he will either roll over on to his back, or cower in a submissive posture, providing you with a golden opportunity to lift him on to all fours and, at the same time, ask him to get on his feet and stand. Most dogs quickly learn to enjoy being groomed, and therefore 'rise to the occasion'!

Left *The Fetch. Bowmont Water Trial, 1981* (Frank Moyes).

A dog which repeatedly flops to the ground when working must be approached by his trainer who, in an encouraging tone of voice, should ask him to 'walk up'. The moment he rises to his feet, he must be asked to 'stand'. This lesson will need repeating often and the dog greatly praised when he complies. It will be helpful in getting the dog to understand what you want of him if, occasionally, you walk quietly over to him and lift him on to his feet. Once the dog realises what the command 'on your feet' means, all that will be required to keep him there will be a stop whistle, quickly followed by the command 'feet'. Eventually he will halt automatically on them without being told.

The do's and don'ts when training a young dog are 'in a nutshell' as follows:

1 Do not interfere with a young dog's outrun until the habit of bringing sheep to your foot is firmly established.

2 Do allow the dog, whenever possible, to lift and fetch sheep without commands.

3 Do teach him to 'take time', rather than repeatedly stopping and starting him.

4 A young dog should not be sent for sheep which are out of sight or far away in his early training. Longer gathers must be achieved by you and the dog positioning the sheep, so that he knows exactly where to find them.

5 Gathering a large flock of spread sheep regularly should help prevent the habit of 'stopping short' developing. (Sending a dog regularly for a small group of positioned sheep encourages this habit!)

6 Before sending a dog to gather, get him to stand beside you and ask him: 'Can you see sheep?'. If he looks in the wrong direction, or at sheep over the dyke, tell him 'No' in a firm voice. This command is so simple to teach a puppy, and extremely useful on numerous occasions when he is trained. A sound young dog should see sheep immediately. However, some dogs, either due to excitement, low intelligence or poor vision, do not.

7 Should a young dog encounter difficulty in moving awkward sheep, always be ready to lend him a hand.

Teaching......a dog......to stand.

'What does "on your feet" mean?'

Flanking

It is insulting a young dog's intelligence to take him to a flock of stationary sheep and insist that first he flanks around them one way, and then the other. It is relatively simple for the trainer to move the sheep first either clockwise, or anti-clockwise, so that the dog's natural instincts take over, and he has in his mind that flanking is done for a reason. You will then be able to fit in the appropriate commands with what he is doing.

There is nothing to equal a flock of wild Cheviot hoggs on a steep and undulating hillside to encourage a young dog to flank freely. (If he doesn't flank, he will most certainly lose the sheep!) Unfortunately not all of us enjoy such ideal conditions and thus must resort to other available means, creating situations which give the dog every opportunity to respond in a natural fashion.

If you insist that a young dog runs around a stationary flock, purely for the purpose of learning his 'sides' and 'freeing him up' more serious problems will be created. The dog will have no idea where the 'point of balance' is on a flock, and this will cause him not only to stop short, but also to stop anywhere except where he should, at 12 o'clock or the point of balance. He will also 'overflank' when running to turn sheep. Only on rare occasions, and with a very determined dog, should we insist that he flanks in a circle around stationary sheep. This exercise would not be necessary if shepherds and farmers who are also trial enthusiasts, kept fewer dogs and thus gave them more work.

The most efficient method, I have found of introducing a young dog to the art of 'shedding', 'driving' and 'flanking' is to go 'walkabout'! The sheep must first

Above *Take the dog by surprise and flank him off the heavy side* (Frank Moyes). **Below** *Encouraging the left flank* (Frank Moyes).

of all be 'shed' from the flock, and driven to the nearest gate in order to take them for a walk! At first these feats must be accomplished more by the trainers' skill than that of his dog, who must be positioned, and told to 'stand' on the opposite side of the flock. The trainer must then calmly cut off half a dozen sheep. With this accomplished, and a suitable space made between the two lots of sheep, he must encourage his dog to come to him (on bended knee if necessary!) The man and dog must then attempt to drive the 'shed sheep' to the nearest gate.

A 'natural line dog' should not find this task unduly difficult as this type is often a good driver without being taught. If the 'shed sheep' do attempt to break back to the others, the trainer must quickly move on ahead, out of the dog's way, leaving him to bring the sheep on his own. His instinct to 'fetch' is often far stronger than his driving instinct. At this point, the trainer will need to encourage and 'lay the dog on' with his voice. The dog should respond with natural skill and enthusiasm. If, by chance, the sheep do manage to get back to the others, so much the better. It will encourage the dog to do everything in his power to ensure it does not happen next time. Once you manage to get the shed sheep through the gate and out of the field, it is of no consequence where you take them. Uphill, downdale, through gates, over streams—the world is your oyster!

Whilst they are fresh, your sheep will repeatedly endeavour to break away. Your dog's natural instinct will encourage him to head them, giving you the opportunity to fit in both whistled, and spoken flank commands. Going 'walkabout' benefits the young dogs in more ways than one. Dogs which will gather equally well off either hand are few and far between. Most, just like humans, have a preference and, to make life more complicated, they can, and do, alternate between left and right.

Usually, the dog flanks happily and at the correct distance from the sheep, on the side he prefers. Eventually he must be taught to flank both ways correctly. Should you become impatient and insist that he flank correctly on the side he dislikes, there is every possibility that he will become sour on that flank for the rest of his days. The dog that flanks freely off both hands obviously does not present any problems, but still will benefit by occasionally being taken by surprise and asked to flank off the heavy side of the sheep and in the opposite direction.

It is almost inevitable that at some stage of training (and often unintentionally) you are going to offend your dog (some dogs are more sensitive than others). When the dog does 'take the huff', he should be kept right away from sheep for a few days until his memory dims concerning the incident.

When, once again, you take him to sheep, do not speak to the dog so that he can naturally and easily perform the task in question on his own. For example, if the dog shows a reluctance to flank to his left, the trainer should first of all position a few sheep alongside a fence or wall. The trainer should then ask the dog to 'stand', thus allowing the sheep to 'break' along the side of the fence. The dog will become excited and unable to resist the urge to head the sheep. The exercise must be repeated all the way around the perimeter of the field, after which he should have lost any reluctance for the flank in question. Should the dog choose to flank too tightly, he must not be interfered with as this minor fault can be corrected at a later date.

When you are sure that the dog is 'sweet' again and into the habit of flanking occasionally, give him the appropriate command; alternating from word to

Above left and below *Walkabout* (Frank Moyes).

whistle helps to keep him happy. The next step, if he is flanking too tightly, is to walk quietly in between him and the sheep and gently push him 'off the corners' with a wave of your stick. After a number of 'going walkabouts' and allowing the dog to do his own thing, call him to your foot, the sheep will surge ahead and the dog can be sent to retrieve them. Because he will be keen and excited by this procedure he may set off 'straight', hot on their heels! If he does, you must first stop him, allow thinking time and then set him off with the long slow 'opening out' whistle discussed previously. Repeat the exercise until he 'gets the message!' (if it is necessary).

After walking some distance, the sheep will begin to tire, making it possible for the dog to hold them to your heels. You can assist in keeping them there by gently moving your stick from side to side in front of you. Holding the sheep to your heels is beneficial in that it replaces the 'balance' which the dog has lost owing to the repeated flanking he has done to retrieve the sheep. Certain aspects of training remove qualities you may wish to retain, therefore means must be devised whereby these qualities are put back again.

Most of what I have written applies to the determined dog. You may own a free-flanking dog with plenty of power which is amenable, and quickly learns his sides (flanking commands) after only two or three lessons and, although it is possible to train a dog (but not to polish it as well!) within the space of three months, others can take up to one year or even longer. Weak dogs, lacking in courage, are generally more than delighted to flank, as it supplies them with an opportunity to leave the 'heavy side' of the flock.

Once you have got your young dog gathering reasonably well, any 'course crossing' must be discouraged. When it occurs, call the dog back to your foot, and hold a conversation with him, asking him what he thinks he is doing. (Friendly chats work wonders!) If he is half way up the field before he attempts 'the cross' you must stop him, ask him what he's doing and then redirect him with a long slow opening out whistle, only speaking gruffly if, after several attempts, the dog continues to 'cross'.

To date I have only owned two dogs which never attempted to cross their course, and one of them surprised me on one occasion by making to come in between the sheep and me. I was disappointed in him until the reason was pointed out to me. The sheep had left the post and were coming full gallop in my direction long before the dog reached them. He saw what was happening and made to 'come in' to slow them down.

Another simple way to teach a dog his flank commands and, at the same time introduce a 'Gettt' command to a 'strong' dog, is to erect a pen. Some dogs are natural penners of sheep, knowing instinctively where to be at the right time. Once you have had the pleasure of owning a capable penner, it teaches you where to position other dogs in order to encourage the sheep to walk inside. For the purpose of teaching him his left and right flanks, the sheep must be allowed occasionally to break around the pen!

When you provide a pen for the purpose of introducing a young dog to the 'art' of penning, it is advisable to use sheep which have been penned already so that the dog will not become despondent after having a difficult time persuading the sheep to go in. A young dog with little or no experience will often try to hold the sheep to his trainer rather than drive them into the pen. He does not know where you want them. To show him, it is sometimes beneficial if you stand outside, at the back of the pen.

Previously penned sheep will soon show the dog what is expected, as soon as they walk in the young dog should be encouraged to stand in the gateway for a few seconds and told how brilliant he is. The gate should then be swung shut. When opened, the dog must be told to 'stand' before permission is granted for him to turn them out, and stand again, so as to allow the sheep to settle before he gathers them up ready for the 'shed'.

Balance

Natural balance, as well as being a great asset, is a rare and much admired quality in a trial dog. A dog fortunate enough to be endowed with balance never upsets the sheep, and can hold them to you with little or no flanking at all. Balance is an essential feature at both the pen and the shed; a dog without any must, as we say, be 'balanced up'.

Encouraging a dog to hold four sheep against a wall is one way of balancing him up. Keep asking him to walk up until the sheep split and break in all directions. After a number of 'repeat performances' the dog will instinctively be in the right place at the right time to prevent the sheep breaking. Once the young dog has become proficient in the art of 'holding sheep to a wall' you can proceed a stage further. You will need to assist him to shed three sheep from the flock and drive them to the centre of the field.

Balancing up.

The dog should be positioned so that he is standing with his back to the flock. The trainer must then gently try to push the three sheep past the dog, swaying first one way and then the other, rather like a charmed snake! Caution must be practised on the part of the dog, who must not overflank when counteracting the sheep's movements. Should he apply too much pressure, the sheep will endeavour to run past the trainer's outstretched arms. When this happens the dog should be quietly admonished, so that he realises his mistake and eases back. Eventually (probably after several lessons) you will be working in unison with a beautifully balanced dog which counteracts both you and the sheep's every move.

Shedding

Some dogs are born shedders, requiring little or no encouragement. To the hill shepherd and in the lambing field a good shedding dog is worth his weight in gold! When competing in a sheepdog trial, and after having a 'good course' and pen it is always reassuring to know that the dog is capable of a 'good finish'.

Shedding is a natural instinct just waiting to be aroused which can only be soured by man's impatience. The dog is only reluctant when he is unsure what is expected of him. Providing you have done your groundwork correctly, by regularly encouraging the dog to come quickly to your foot from puppyhood, it should not be unduly difficult to get him to come to you through the middle of a quiet flock.

The first step is to get the dog to stand on the opposite side of the sheep, allowing half of them to move well away. When there is a sufficiently wide gap, call the dog to you. He may show no hesitation or, on the other hand, he may look completely bewildered! His instinct will tell him to flank round the

Above *Persuade your dog to come directly to you.* **Below** *The Shed. The Bowmont Water Trial, 1981* (Frank Moyes).

Jan working the 'single'.

perimeter of the whole flock, and put them all back together again. You must persuade him (again on bended knee, should it be necessary) that it is much more important that he comes directly to you. Eventually you will convince him and he will come, probably wearing a sheepish expression! Be lavish with your praise before encouraging him to assist you in driving the shed sheep well away from the others, this being the easiest method I know of to teach a young dog to drive.

The line dog will become a proficient driver almost at once. The flanking sort usually endeavours to head the sheep at every given opportunity (as does the weaver). The best method of counteracting this move is to head them quickly yourself, encouraging the dog to fetch them after you. Eventually, when the habit of coming through the flock is formed, you can narrow the gap gradually until the dog will run through the flock without one.

Like everything else, teaching the dog to 'shed' will not be accomplished overnight. Always make sure the sun is not in the dog's eyes before asking him to 'come in'. Gradually decrease the number of sheep which the dog sheds off. Shed regularly from both ends of the flock; a one-way shedder uses up too many valuable minutes at a trial.

Excitement should be conveyed to the dog by the tone of your voice when you ask him to come in. I say: 'Here, this' when more than one sheep is required, and 'This one' when he has become proficient, and I want a 'single'. Once the dog has safely shed his sheep, lay him on, encourage him to take charge so that not only will he be confident in what he is doing, he will also enjoy doing it!

Take the shed sheep 'walkabout'. As soon as the dog gains confidence and begins to take the initiative, allow him to push on ahead, with you gradually falling back, encouraging him to drive the sheep as far as he will on his own, occasionally through gaps and gates, over bridges, up steep hillsides, until he is completely in charge.

A useful exercise for putting the dog 'on his mettle' is to shed a few sheep from the flock and drive them a sensible distance from the rest. Ask the dog to stand, allowing the shed sheep to bolt back in the direction of the others. Before they reach their destination, send the dog to 'head them'. You will need to call: 'Here, this' to get him to come in. It is a beneficial exercise which most dogs, providing they are fresh, thoroughly enjoy!

Often when competing at a trial, sheep are extremely difficult to shed. There never seems to be a suitable gap for the dog to come through. You ask him in, and then have to tell him to get back out again. By the time he does get the opportunity he is often thoroughly confused; and no wonder, who wouldn't be! Running on four sheep, at some trials, you are told that when you come to the shed, it is a 'split' meaning that it is permissible to take any two sheep. At other trials it is the last two sheep which must be cut off. Towards the end of summer the newly-weaned, and hopefully more sensible, ewes come on the scene, with them it is the last sheep that is required. Poor long-suffering dogs, nobody gives them a thought!

When running on three sheep, and your 'time' quickly disappearing, if you reach the shed it is just as well to bang your dog in, you are bound to end up with a two and a one. It is better to have a rough shed, than no shed at all!

Some young dogs, and older ones, are slow at coming in, so slow, in fact, that they can miss an opportunity. The chances are that if you scold them, they will refuse to come in at all. Eventually, when you have taught the dog his slow and fast flanking whistles, you will be able to encourage him to shed quickly with a whistle. After regularly getting him to shed in this manner he will soon acquire the habit of dashing in at speed.

Much more care is required when attempting to shed with a 'line' dog, as the majority of them are not as proficient as flanking dogs. The reason is that a 'line' dog tends to come into the gap 'straight', (if two sheep are required, and they are travelling at speed, you can end up with one!). So more care is needed, plus a decent sized gap. A flanking dog curves in on to the shed sheep, enabling the handler to be able to take much greater risks. This is very advantageous when time is short (and it looks pretty impressive too!).

The 'Look back for more sheep' command

Once the dog sheds and drives sheep off with enthusiasm, it is time to teach him the 'Look back' command. This command is extremely useful on the hill, and highly essential should you be fortunate enough to be chosen to run your dog at an International.

Start by shedding a few sheep from the flock, and drive them a little way off from the others. Distract the dog's attention away from the shed sheep by calling his name, encouraging him at the same time to look in the direction of the remaining sheep. To do this you may have to walk a few paces with the dog, in their direction. Ask him if he can see them, when he does, say clearly: 'Look back' and follow it up with the appropriate whistle. He should soon learn what is required of him. Before very long you will be able to stand him in between the sheep he has shed off and the sheep you wish him to 'go back' for.

It is a good idea to stand a young dog on the side of the shed sheep on which you are going to put him back. For example, if you intend sending him on a right-hand outrun for the remaining sheep, flank him to his left on the shed sheep first of all, and then proceed to put him back from that position. An older, experienced dog, can be put back from directly behind the shed sheep simply because he can, should it be necessary, be redirected with an opening out whistle.

On the hill it is imperative that a young dog be allowed to fetch the first lot of sheep he gathers, right to your foot, before being put back for those which suddenly appear out of nowhere (and those which he has inadvertently missed). The reason is that an impressionable young dog will start 'going back' without being asked, and even when there are no more sheep to go for. I have seen young dogs setting off to go back (leaving the sheep they have gathered literally 'out on a limb') with nothing more than the drop of a hat, or a stop whistle as the signal, simply because they enjoyed doing it, and wish to please or occasionally they are fed up with the sheep in their possession!

If you regularly allow him to fetch the first cut of sheep that he gathers, right to your foot, this annoying habit will not be allowed to develop. Once the dog becomes fairly competent at going back, gradually increase the distance that you send him, and discard 'Look back' for the appropriate whistled signal. (Mine is a straight note followed by a curved one.)

At a double lift sheepdog trial, where two lots of wild sheep are to be gathered, it is often difficult to get the dog's attention away from his first 'packet' of sheep when you want him to 'go back'. It is definitely advisable to practice first at home using wild sheep, prior to a trial of this kind.

When first introducing a young dog to the 'Look back' command, it is vital that you keep calm and do not lose your temper if he appears not to comprehend. Impatience will only cause the dog to become utterly confused, resulting in him 'burling round' with his 'rudder' in the air! or, as some sages put it, 'waving his flag'. Not only does this look dreadful, but it can easily develop into an incurable habit.

Driving

The 'line' dog is very often a natural driver, happy to settle in behind sheep and take them anywhere. Some dogs show a marked reluctance to drive. These dogs have to be taught, using methods that are carefully thought out by the trainer so that the dog grows to enjoy the task. Should it be necessary, the trainer must be prepared, over a period of weeks (or in some cases months), to walk many miles with the dog and sheep.

I have mentioned previously the quickest and easiest way of teaching a dog to drive. This entails shedding a few sheep from a flock and both you and the dog driving them off to a suitable distance. Should the sheep escape back to the main

Gathering and driving sheep.

Incorporate driving into the dog's normal day.

flock, this will only serve to encourage the dog to be more cautious next time. The dog should be encouraged to drive the shed sheep up to a fence or wall. Continue to ask him to walk up. When the sheep finally break, the dog will flank to put them back together. This exercise will encourage the dog to 'line' sheep up on his own, when he is driving them.

When driving with 'weaving types' and dogs which lack 'eye' they tend to slide around the sides of the flock, heading them at every opportunity. Occasionally they must be allowed to do this. Nine times out of ten, when training a young dog, prevention is wiser than the cure. However, in certain circumstances, offending traits must be allowed in order to discourage them. By allowing the young dog to head the sheep, and by flanking him behind them again, you are helping him to understand that you wish him to drive.

Dogs which refuse to settle in behind sheep should be taken to a wall or fence, and be encouraged first to drive them round the field one way, and then the other. The dog must walk behind the sheep, with you walking alongside them, thus covering the open flank. The 'weaving' dog benefits from being kept well back off the sheep, so that when you ask him to 'walk up' he will acquire the habit of doing so 'straight'.

Whenever possible the natural driver must be left to 'do his own thing'. If he pushes the sheep along too quickly, he must be stopped and told to 'Take time'. The trainer should try, if there is a need, to walk slightly in front of the dog to prevent what is known as the glancing habit. There are numerous reasons why a dog every now and then will have a quick glance or even a prolonged stare, in your direction when he should be watching his sheep. Young dogs lacking in confidence do it if the trainer drops too far back. As they grow older and more proficient, they usually 'snap out of it'. Working away from home in new surroundings, the habit tends to disappear.

'Over commanding' by the trainer can also be a cause. A timid dog is reassured by a firm voice when he is young, but should be gradually left to get on with the job himself at every opportunity. It is relatively easy to get into the habit of over commanding, especially when either working a determined or a soft dog. The spectator feels like putting his fingers in his ears. Goodness knows how the poor dog feels.

It makes me feel very defensive when an intelligent 'old age pensioner' dog is criticised for 'glancing'. He does not deserve condemnation and is not, as a rule, lacking in self confidence or losing his concentration but merely asking for guidance. Old dogs, just like elderly people, begin to lose their faculties, namely their hearing and eyesight; it is a gradual process which deserves taking into consideration.

Once the sheep and the young dog have run off their enthusiasm, going 'walkabout' is a great way of improving the driving habit. The dog by that time will know the surrounding countryside pretty well and most probably will take the sheep on ahead, lining them up for gateways on his own. This lesson can easily be incorporated in the dog's everyday work.

A useful exercise to improve a young dog which is inclined to overflank past the point of balance consists of shedding three sheep from a flock and driving them a suitable distance from the others. The trainer should then walk first in a clockwise direction around the remaining sheep and then in an anti-clockwise direction, leaving the dog to hold the shed sheep to him. Should the dog leave the point of balance by over-flanking, the sheep will endeavour to break back to the main flock. It is surprising how quickly the dog improves his flanking technique after losing the sheep a couple of times.

Cross-driving

Once a dog has learned to drive, teaching him to cross-drive presents no problems and he is quickly taught. The trainer should position himself in the centre of the field, allowing the dog to drive the sheep round him in a circle, first one way and then the other, gradually increasing the distance by enlarging the circle. At this stage, you will find your 'Come here to me' whistle extremely useful.

It is very useful to encourage the dog to come towards you when cross-driving, especially when you wish him to move quickly. On the fetch this whistle can be used effectively to draw the dog quickly down the sides of the sheep. It can also be used to pull a young dog off the corners of sheep when turning them in a different direction. Just as soon as the dog comes to realise what is expected of him, the 'Come here' whistle can be dispensed with and substituted by the appropriate flank whistles.

All lessons must be repeated regularly as it is through repetition that the dog becomes proficient. However, there is a happy medium and should the dog appear in any way reluctant or overtired, do rest him up for a few days. He will then come back to his work as fresh as a daisy. Training every other day is quite adequate, with the dog being taken for a walk away from stock on his days off.

Owing to the fact that a 'natural' dog needs very little correction from the trainer, it can take an extra trial season before he comes into his best as some natural dogs work by instinct alone instead of using their brains. A determined dog which pushes sheep harder and works tighter on them requires a great deal more concentrated effort on the part of the trainer. Should a difficult situation arise both man and dog will be able to cope more efficiently because of this extra effort.

A dog with a natural inclination to line up sheep is a useful asset, especially on the cross-drive as he can keep the sheep travelling in a straight line with often only the command to 'Take time'. Cross-driving with a flanking dog is accomplished more by the skill of the handler, who must see to it that the dog is always in the right place at the right time.

When, finally after months of careful training, you can work the dog without taking your eyes off the sheep, when you can trust him to flank just enough, at the correct speed and distance, in tune with the slightest turn of the sheeps' head, when the dog has become an extension of yourself, then you and he are ready to compete in 'open trials'.

Travelling many miles all over the country to sheepdog trials is hectic and exhausting for all concerned. Men and women have a choice which generally children and dogs do not have. The day after a trial or trials, we endeavour to give our dogs a complete rest, whereas GWB Junior, if given the choice, would choose a day of fun and games!

It needs to be mentioned that the training of a keen young dog is a relatively simple task compared with keeping him on form. By the term 'on form', we mean receptive, enthusiastic and up to trial standard throughout his working life. There is only one way to accomplish this and that is to treat him as you yourself expect to be treated; mainly to ensure that he receives the necessary nutriment, has a warm and comfortable bed and is not worked off his feet, bearing in mind that all work and no play makes 'Jack' a dull boy!

Penning

There is nothing more soul destroying for a young dog than to be flanked round and round a pen when there is no hope of ever putting the sheep inside. Except when running a dog at a 'national' or in a team event, a wise handler in such circumstances would put his sheep away rather than persist at the pen. There are a number of reasons why sheep refuse to be penned. The main reasons are: the pen is too small, the sheep are in a panic due to the dog's abuse around the course, often the sheep are badly shepherded on their home ground and, occasionally, the handler, instead of standing back out of the way, insists on standing too close to the mouth of the pen. (One of my numerous mistakes!)

Once in the vicinity of the pen mouth, wild sheep require lulling into a feeling of security. This can only be achieved by a well balanced dog turning right off them each time they attempt to escape. As soon as the sheep come to realise that the dog does not mean them any harm, but has no intention of allowing them to escape, they will stand and watch him. This is the moment to ease the dog up quietly to them. Should the sheep break yet again, the dog must immediately give ground without allowing them to escape until they realise that the wisest alternative route is to go inside the pen!

Penning with a particularly determined dog can, and does, present problems. He will probably give you a great course until he gets the sheep near the pen mouth, where he will promptly disregard any instructions you care to impart, endeavouring to put the sheep inside entirely on his own with a couple of quick, tight turns! Although the dog has undoubtedly ruined your run, you cannot help but admire his independence, at the same time blaming yourself for not having done your homework correctly.

Usually the result of the dog's bombardment is sheep scattering in all directions! Blackfaces (which are, as a rule, quiet, heavy ewes) and Lowland hoggs are the only sheep to tolerate this form of abuse. When working gimmers, Half-breds or Mules, nine times out of ten you will have a break!

From an early age, the determined dog should receive *extra* home tuition at penning. He must be taught to throw himself back off the sheep, only approaching when asked, and with extreme caution. The command to 'Gettt' must be firmly endorsed and obeyed without hesitation.

Only once was I fortunate to meet the late J.M. Wilson. He told me that a sheep's eyes dilate when they are about to walk into the pen. (Unfortunately I am unable to find out whether there is any truth in this statement as Garry never gives me the opportunity!) I have, however, seen bird's eyes dilate when they

Loch and I penning sheep at Peebles.

are spoken to in a soothing tone (and in broad daylight!). Also I have mesmerised sheep in this way and been able to catch them without a dog. The secret is never to look at them, just to keep talking quietly and inching up until finally you pounce, breaking the trance. (If you miss, don't count on a second chance!)

Some sheepdog handlers are masters in the art of penning sheep, often winning the trial at the pen. Few men are rated in this category. I was fortunate to be seated beside one of the best, at Kilmartin Scottish National a few years ago. The sheep on that particular day were Blackface gimmers and, on this occasion, particularly troublesome at the pen. I could hear 'the expert' talking quietly to himself. Curious to hear what he was saying, I leaned closer. He was studying the dog at the pen and whispering: 'Lie down, lie down'. The dog obligingly lay down and the sheep walked confidently into the pen.

Half an hour later, Garry and I stepped out on to the lush green field, the sun's reflection dazzling on umpteen windscreens down the east side of the course. We had a useful 'run' until the later stages. After missing the last 'gates' and completing the shedding we proceeded to the pen where I promptly ignored the advice of the 'expert'. 'Get on to your feet', I said (showing my lack of sheep sense). Needless to say, I never got a second chance at that pen! (The 'experts's' name? Jim Gilchrist of course!)

Nursery trials

There are many valid arguments both for and against the running of young dogs in Nursery trials. Some say 'nurseries' ruin a young dog. Certainly training a dog in such a way as to expect him to win will dampen a young dog's enthusiasm. I have seen many heartily sick by the last Nursery of the season, and a high percentage are never to be seen again on a trial field.

However, providing the handler takes the right attitude, which should be to use the facilities provided for the benefit of the dog and not for the benefit of himself, ie, to give the dog confidence and valuable experience to prepare him to run in Open trials, I cannot see much harm in Nursery trials. Local Nurseries in Scotland allow dogs up to the age of 2½ years on the first day of October, to compete. Commencing in October, weather permitting, there are usually several trials held at various venues throughout the region. At the end of the season, the five highest pointed dogs from each region go forward to the Nursery Final, held in March.

In order that the young dogs be given a confident introduction into the art of trialling, a great deal of thought must go into the construction of the courses. A good idea is to place a bale of sweet hay beside the 'Lifting post' (which ideally should be painted white). This encourages the sheep to stay put without the attendance of either man or dog. The less the young dog is distracted the better. The letting out pen should be well hidden from the dog's view and the outrun should not exceed 250 yards, gradually increasing the distance of the outrun as the season advances.

When competing in his first trials, an intelligent young dog can, on being asked if he can see his sheep, become agitated or confused. The reason is that (of course!) he's seen them and thinks that there must be others that he cannot see. Should this situation occur, try to set him off with the appropriate flank, not

Nursery trials? The long and the short of it.

asking if he can see the sheep.

At local Nursery trials fetch gates are dispensed with to enable the dogs to bring and line up their sheep with as few commands as possible. I feel that fetch gates are unnecessary at any trial. Granted a dog must flank to put sheep together on a hill, although a capable dog should be able to achieve this virtually unaided, as he will be much happier doing it for a valid reason. Throwing a dog about, here, there and everywhere, when he is attempting to bring sheep to you will not only go against his natural instinct but also dampen his initiative and enthusiasm.

The average trial course is far too short for the use of fetch gates to serve any practical purpose. Wild sheep are upset further by the shepherd at the letting out pen being required to hold them to the post—a skill at which few men are adept, even with well behaved sheep. Without 'fetch gates' sheep can be picked up from any position with less cause for concern on the part of the handler.

The 'easy' type of dog would quickly become more proficient through not being over-commanded in the initial stages of a 'run'. A handler whose views I normally respect, except on this issue, argues that a dog is expected to flank on command when fetching sheep over a distance and driving them through gates at home. He misses the point completely. A dog with any brains at all (if he is allowed to use them) quickly learns where all the gates are situated and is quite

Geoff Jr herding ducks.

capable of bringing the sheep through them on his own—a relatively simple task in fields with a fence at either side of the gates!

From fetch gates we come to the starting post which should be positioned, preferably, well to the fore of the judges' car so as to allow plenty of room to turn the sheep behind the handler. A large pen is provided, to put the sheep in and sufficient time allowed in the hope that every young dog has a good chance of penning and completing the course.

At the shed any two sheep (regardless of which way they are facing) can be taken off. Much to the amusement of spectators and fellow-handlers this sometimes results in the handler shedding and walking away in one direction whilst the dog drives the 'last' two sheep in the other!

It is nigh impossible to judge a young dog's merits on one showing. Consistency over a long period is what counts. At a trial you will hear spectators state emphatically: 'That is the best young dog I have ever seen!' By the following event they will probably have had a complete change of heart and be making exactly the same statement about a completely different canine. There is only one person who really knows the dog and that person is his owner. In the long run it is often the dog which suffers the most criticism which ends up the best of the bunch.

If you own a dog which is being discussed by the media, favourably or otherwise, it is often a useful dog. It is not what a young dog is capable of achieving that is important; it is the way that he goes about it! In certain circumstances, it is wiser to allow offending traits so that you are given the opportunity, eventually, to discourage them. We should all realise that if we are prepared to learn, every dog that we own will teach us something new and beneficial in the training and handling of our future dogs.

Chapter 12

Sheepdog trials

A fear of guns
Even though most Border collies are reared in a similar environment, a small percentage inevitably grow up to be gun shy. This can put the handler at a distinct disadvantage when working his dog at a trial where sports or a clay pigeon shoot are going on.

The dogs' temperament does not appear to have any bearing whatsoever on whether or not he will have a fear of gunfire and similar loud reports. We have owned both timid and bold dogs which were gun shy. Jan would tremble with fear, and was so afraid of gunfire that she had to be kept inside the car with the cassette playing at full volume (and not the *1812 Overture!*).

A couple of years ago, whilst attending a sheepdog trial at South Shields on the rugged east coast, a mine exploded out in the North Sea, not far from the shoreline during Jan's run. After the loud report the ground shook. Instinctively Jan, myself and the least brave of the numerous spectators flattened ourselves on the ground in fear! Although courageous little Jan stayed with her charges, it was impossible to salvage the run which, up to that stage, was clearly the winner. The same thing happened at Yetholm Border Shepherds' Show, only this time it was the sports starting pistol which proved to be the culprit and forced Jan into second place.

Car travel

I have known many dogs which refuse point blank to travel within the confines of a car, and yet the same dogs would leap joyfully inside any open boot, obviously preferring the quietness and darkness to the sight and sound of rumbling traffic. Providing the dog is happy and quite contented in the boot of a car, and there is adequate ventilation, I do not consider this mode of travel to be at all cruel. On the other hand, enclosing a dog on a boiling hot day in a car boot, for hours on end, is not only terribly cruel but also foolhardy. There is no doubt that Border collies have been known to perish given this kind of treatment, leaving behind an owner much wiser, unfortunately after the event!

We always travel in an estate car, mainly for convenience as it can be used for more than just carting canines! A white car is preferable as white deflects the sun. When purchasing a car with carting dogs in view, it is wise to buy one with a perpendicular rear window. Hot sunshine beating down on dogs through a sloping window must be like riding in an oven! It is surprising the number of estate car designers who do not take this into consideration, often ensuring that sensible dog owners purchase a different make and model to the one they prefer.

It is always advisable when parking the car to do so in the shade. If others have got there first, a rug draped over the window will suffice. A 'hot dog' should only be given a small amount of water after a run; once he has cooled down he can then be allowed to quench his thirst. It is inadvisable to give him water before a run, for obvious reasons!

Some tips for the trials

Trial sheep which are handled quietly with care and consideration within the letting out pen often behave reasonably well on the course compared with sheep which are handled roughly and noisily. It is when the letting out pen begins to empty of sheep that the few remaining inevitably become upset and suspicious, causing the rot to set in.

Rain often settles sheep, whereas wind stirs them up. A combination of hot sun and annoying flies will send them scurrying for the shade! Often the first and last sheep of the day are inclined to be the easiest to handle. There is an exception to this rule, in the shape of gimmers who (so I've been lead to believe) become frisky of an evening!

On an average-sized trial course it helps the dog to see the sheep and gather them correctly if both he and his handler walk on to the course in a direct line with the starting post, and from behind the judge's car. On a big International-type course it is helpful to the dog if he is walked on at a slant from the opposite side from where you plan to send him for the sheep. This way his nose should be pointing in the right direction for a wide outrun.

It is always common sense to study a trial course carefully, and to see as many runs as possible before you run in order to watch sheep's behaviour, and tactics (especially those of the 'top' handlers) although mimicking them is often another kettle of fish! More often than not less points will be deducted by an efficient judge for a missed set of gates than would be for sheep which are chased and fought in order to put them through.

Recently, in Sheila Grew's marvellous little magazine *Working Sheep Dog News*, there appeared a letter from a Swedish trialist seeking advice on how to cure his nerves because they had a serious effect on his ability when handling his dog amongst other competitors. Psychologists tell us the reflexes of a slightly nervous person are much sharper than those of someone who is in a relaxed frame of mind. However, a handler who is absolutely petrified is bound to transfer his fear to the dog who, I believe, due to his primitive instincts, can actually smell the agitation his master feels and thus reacts unfavourably. Espinar said in 1644: 'Partridges lie much better to the dog that finds them not by foot-scent but by body scent, and measures his distance by their tameness or wildness, for he can tell by the body-scent if they are restless or tranquil.'

Running a dog in a team event is always nerve-racking, owing to the fact that there is always the worry that you may let your side down. When running for Scotland, and doing badly, I attempted to walk off the course, but was hurriedly chased back by irate team mates. Because I was dreadfully nervous on this occasion, I completely forgot that my points, no matter how few, were still necessary.

I have found that running a strong determined dog, once you are actually out on the course, will often keep you so busy that you tend to forget your nerves. Working a slack, easily-handled type of dog, you become all too aware of what is going on, and what is being said around you. I believe that what helps nervous handlers most is that if they fail to pull themselves together they inadvertently fail their dogs and they must tell themselves just that, prior to competing.

When and if you ever receive compliments about your dog, or handling ability, it is wise to remember that within the trialling fraternity you can be discussed and have your praises sung from John o' Groats to Land's End one season, and be completely forgotten by the next! It is very easy to come to believe that your dog's faults are practically non-existent, but very foolish considering there is no such thing as the perfect anything. All dogs have faults—you hope that, with age and experience, these will tend to lessen as do the faults of their handlers. The sensitive dog develops more confidence, the weak dog acquires adequate power to get by and the very determined dog (providing he has a patient handler) will gain enough sheep sense nine times out of ten.

When dogs run away

During the sheep dog trial season, after spending most of his day in the liquid refreshment tent, an inebriated shepherd will, on occasion, be carted off home minus his dog. One such 'gentleman' returned the following day to find 'man's best friend' patiently awaiting his return. This shepherd counted himself fortunate in owning such a wise and faithful companion. (The dog, no doubt, had seen it all before!)

Some dogs, when lost or abandoned, will set off for home, nearly always striking out in a wide circle. (How long it takes them to return to the spot they set out from depends on how far they have travelled or how wide a runner they are!) If two or more dogs join forces there is always the possibility that they will worry sheep, especially in the case of young dogs or if a terrier is involved.

A lone lost dog will occasionally pick up a few sheep in all innocence and, whilst endeavouring to take them to his master, he may meet up with an obstruction in the shape of a stone dyke or fence. It stands to reason that, in his perseverance to keep the sheep on the move, he may become excited and start worrying, eventually maiming or killing some of them.

If a dog goes missing, his owner must do everything in his power to discover his whereabouts as quickly as possible. A number of years ago a farmer's two young Border collies escaped and worried some free range cockerels which were being fattened up for Christmas. When their master discovered the damage, he rushed into the house for his gun. Fortunately for the dogs the telephone rang and he was kept in conversation for some 15 minutes. By that time the irate farmer had calmed down and decided to postpone the execution until the following day. He never did get around to shooting Bill and Ben, blaming himself for their 'escape'. In return they turned out to be the greatest working dogs he ever owned!

Chapter 13

Two dogs of note

Having described at length how to go about training the ideal Border collie, I would now like to mention two dogs in particular.

Don

In February 1930, Christopher Graham, farmer of Nab Hill in the county of Northumberland, lent Don, a stylish, rough-coated collie, to his brother, Bill, who was then shepherding beyond Edinburgh on the north side of the Firth of Forth. Bill Graham travelled with Don by train, to his home and, thinking that the dog would settle, released him. However, being a collie of rare and exceptional intelligence, Don wasted no time in making tracks back to his beloved master, Christopher, at Nab Hill, some 70 miles to the south.

Don, a collie belonging to Mr C. Graham, Nab Hill Farm, on the Northumberland border, which trotted home five days after being sold to a man in Edinburgh. The distance is over 70 miles.

Don, who walked over 70 miles back to his owner.

To this day no one knows whether, on reaching the Firth, the great dog took to the icy waters, swimming against the strong and dangerous current, or whether he was fortunate enough to find a less arduous way across one of the bridges. After five long days and cold nights this faithful and truly courageous collie, weary, hungry and footsore, wandered into the yard at Nab Hill to a tumultuous welcome, where he was promptly declared a local celebrity and his story portrayed in both local and National newspapers. It goes without saying that Don and Christopher became inseparable companions until Don died peacefully at a ripe old age in the home he loved.

Right *My aunt, Sister Lucy.*

Below...*and her namesake.*

Lucy

At the time of writing, Lucy is the latest addition to our ever-growing family of dogs. There are many amusing aspects of her character, making it difficult to know how to describe her with any real accuracy. Mainly due to her over-enthusiastic nature (and that is putting it mildly!) I have already begun to train her for work, although she is only eight months old. Providing I do not rush her or cramp her particular brand of style, and bearing in mind that the average bitch matures earlier than a dog, I hope to have her going reasonably well by 'the lambing'. I must admit to being curious as to how a pup with her funny little peculiarities will turn out.

You may wonder as to why I christened this bundle of mischievious vitality

'Lucy'. Well, I have several valid reasons, mostly in the shape of a loveable and sweet-natured aunt of the same name. She is my mother's sister, and the youngest of a family of five children (three girls and two boys) born to my maternal grandmother—another Lucy! There are most definitely similarities between Aunt Lucy and the four-legged one. Mainly that they both possess a delightful sense of fun, are constantly on the go, helping where they can (and consequently both have a healthy appetite!). I might also mysteriously add that they are both beautifully marked, with a traditional full white collar!

Aunt Lucy loves country pursuits and thoroughly enjoys her visits whenever her 'occupation' (she is a nun) allows. Children and puppies alike especially adore her company. Puppies especially find her irresistible, they follow her everywhere. In their eyes nothing and nobody else exists when she is around. Her explanation for this being that they consider her to resemble a 'mother sheepdog'!

Lucy's work invariably takes her overseas, however she often manages to spend some time with us here at Swindon in the summer months. 'Her puppies' are always dying to make her acquaintance!

Puppy Lucy conveniently made her appearance on the scene the day before Penton Sheepdog Trial. I say conveniently because we were luckily able to kill two birds with one stone by being able to run our dogs and visit the 'new' puppies on the same day. There were five in number—four bonny bitches and a nicely marked dog. Long before the puppies were born I had pictured Lucy in my mind's eye, and subsequently had no difficulty in deciding which puppy to choose.

Midge, Lucy's mother, is a smart prick-eared bitch with a lovely deep muzzle. Glen is Midge's sire and he is a litter brother to our Jed and Trim. Midge's dam was Jean, an almost black bitch. Sadly she died last year, having been a tremendous worker and living to a ripe old age. She too had a great deal of Bob Fraser's 'Mindrum' blood coursing through her veins (plus a dash of Welsh!)

Originally we had hoped to mate Midge with Garry but he, true to form, was feigning one of his 'headaches', and so the chore of fathering Midge's offspring was handed over to his son the 'ready, willing and extremely able' Loch, who, nine weeks to the day, was aptly rewarded by the fruits of his labour! The dog puppy remained with the breeder and the three bitches went to farm homes, while Lucy took over at Swindon! By three months of age all Midge's litter had begun to work.

By seven weeks of age 'our' Lucy was fluffy, fat and fun loving. She was also extremely pretty with a traditional full white collar and brow, white chest and forelegs and there is a touch of fawn on her cheeks and hocks. A rather unusual feature, which has all but disappeared, is a scattering of silver hairs mingling with the black on her hind quarters. Her coat is exceptionally silky and dense, not unlike that of her great, great grandmother, Meg. Lucy's ears, large dangling appendages, were her worst feature when she was a puppy. They really were gi-normous, rather resembling those of the cartoon character 'Dumbo', the baby elephant. Thankfully it did not take her too long before she grew into them.

Lucy, to my mind, is just a fraction 'close bred'. Line breeding when the blood on both sides is similar four or five generations back works wonderfully if the ancestors are in themselves sound, but 'in-breeding' often results in a 'hot' excitable temperament appearing in most of the offspring. Our idea was to produce a breeding bitch with two crosses of our 'old Meg' in her pedigree and this, thanks to Loch and Midge, we managed to achieve.

Lucy answered happily to her name on arrival. Car rides are her idea of heaven, the longer the journey the wider her smile! (Although no persuasion is required to get her into the car, a great deal of patience is required to persuade her to come out!) Inoffensive ducks, squawking hens, even cats—in fact anything which so much as moves—are stylishly and incessantly worked by Lucy, and woe betide any creature which attempts to make a run for cover.

At the tender age of 11 weeks, Lucy had no bother whatsoever catching and turning a 70 lb sheep upside down, with incredible neatness and efficiency (an extremely useful feat when lambing flighty Cheviot gimmers, undoubtedly inherited from great grandmother, Jed).

On washdays, Lucy is shut in her kennel—the sight of clothes constantly on the move, revolving in the automatic washing machine, tantalises her beyond endurance. She stands in a 'classical pose', eyeing the machine, before proceeding to stalk it from various angles until, finally unable to contain her exuberance and bent on destruction, she charges it, and collides into the glass door with what can only be described as a biff, a bang and a wallop! Therefore, worried about her safety, and my sanity, I now have to banish her from the kitchen—but only on washdays!

During any spare moments Lucy will also 'eye' the hen house, standing with

Lucy.

Lucy, doing her own thing!

raised forepaw, hoping against hope that a hen will deign to pass by the window, showing its shadowy silhouette against the dusty frosted glass. Whilst I endeavour to unfasten the frayed string (all that prevents Brer Fox from devouring the ducks) often Lucy will tug at it impatiently, anxious to set eyes on the occupants.

'We' regularly pay a visit to a steady flock of ewes. Little Lucy likes nothing better than to get tucked in behind them, often 'driving' some distance on her own. Much to my relief she will run around the flock occasionally. A dog which will only attempt to drive is often more difficult to teach to gather. Sometimes, she will chase and grip the sheep which calls for gentle chiding as I do not wish to offend one so young, believing that eventually the habit will leave her.

No doubt, like other dogs, Lucy is bound to have her faults. Some raise their heads quite early, others lie dormant until a suitable occasion arises. Whatever her faults, it takes all sorts to make a world. (That also is my excuse and I'm sticking to it!) Unlike many youngsters, Lucy is never any trouble at bedtime. She always trots willingly into her kennel (presumably more than ready to count Cheviots?). For a few moments she will stand surveying me with a questioning expression until I bid her a fond goodnight, and quietly close the door.

Chapter 14

Sheep—what would we do without them?

I would like to take this opportunity to pay much deserved homage to the humble sheep. We owe these timid, unoffending creatures much more than is generally realised and are, on occasions, guilty of misusing them. Perhaps we should count our blessings, bearing in mind that without them the shepherd would become obsolete, the Border collie would be put to uses other than those which it was originally bred for and sheepdog trials would be a thing of the past. After shepherding both in-bye and out on the hill, I have no doubt whatever in my mind as to which type of 'herding' is easiest for man and least stressful for sheep.

It is on the hills and open plains that sheep, and man alike, find peace and contentment. It is a natural habitat with rock-strewn acres of bracken, heather, huge tufts of grass and hollows to lie snug and warm in whilst sheltering from the storm. Yet man will persist in going against nature, imprisoning the poor sheep in cold, bleak fields, and on wet and muddy turnip breaks, with little or no shelter at all, causing him a great deal of work and them a great deal of misery.

The sheep's natural instincts warn it, come nightfall, to make for the highest vantage point where it can lie contented under the stars on dry ground, safe from the predators which long ago roamed these lands. At dawn, depending on the

weather, it will stir itself, drifting down to the lower ground throughout the day to where the sweeter grasses grow.

Hill sheep observed from above will often be seen to rest and graze in a 'safe' horseshoe formation so that they are in a position to see the approach of predators—'old habits die hard'. I have witnessed often the tender love and care which a ewe lavishes with gentle mutterings upon her new-born lamb, and felt an inadequacy within myself as a human mother, and on more than one occasion I have stood by helplessly whilst a dying ewe, too feeble to rise, used what meagre strength she possessed, to raise a hind leg so that her precious offspring could suckle the few remaining drops of milk from a slack and withered udder.

By nature sheep are gentle creatures whose only wish is to be left in peace, to congregate and graze together in family groups, which can include great-grand-mothers and various other close relatives! Sheep are thought of, wrongfully, as foolish creatures. This is far from the truth, treated kindly and given time to gather their thoughts, they are capable of showing great wisdom just like any other animal.

Ewes are normally docile but, should a ewe's lamb's safety be in any way threatened she can speedily react, not unlike an angry bull—snorting, stamping and raking both ground and air with a sharp forefoot. When all else fails to ward off the adversary, she will charge, head down, more than willing to sacrifice her own life for that of her lamb. Ewes have also been observed to kill adders with their sharp cloven hooves, crushing them to a pulp if they were considered a threat, either to themselves or their young. Naturally there is always the exception to every rule. In the ewe's case it is usually the 'bad milker' which is the bad mother. Possibly the reason why she will disappear rapidly into the sunset (at the appearance on the scene of shepherd and dog) usually minus her infant, is that she herself has sense enough to realise that there is no hope of providing it with the sustenance it will require.

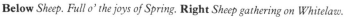

Below *Sheep. Full o' the joys of Spring.* **Right** *Sheep gathering on Whitelaw.*

Wild young hill sheep or 'gimmers', in their first year of motherhood and not as experienced as the older matrons, have to be approached with care otherwise they also are quite capable of 'taking off', usually after first emitting a shrill warning whistle down flared nostrils! There is little to compare with the sinking feeling that a shepherd experiences as he watches a 'stotting' sheep, endeavouring to bluff her way out of what she feels is a dodgy situation, rapidly departing for fresh pastures. Sending an experienced, and therefore reluctant, Border collie in 'hot pursuit' very often only makes her 'stot', or jump, all the faster!

The lambing storm

I will never forget the lambing storm of April 1981. Ernest Crisp always warned me to expect at least two cruel days during lambing; one of which being the 'Pee-sweep' storm, the other the 'Storm of the Curlew' or 'wharp'. The storm of 1981 goes down as the worst in living memory. Within the space of a couple of hours all known landmarks were completely obliterated by a vast mantle of undulating whiteness, which the strong south-easterly wind speedily whipped into deep and perilous drifts.

I remember waking up in the early hours of the morning to a kind of hush and a pure brilliance which lit up what lay beyond the coverlet (which just failed to protect the tip of my nose from the cold air). Still sleepy and only half awake, I slowly became aware of a low moan which only a snow-laden wind sweeping down the valley from the heights of Cheviot, is capable of making as it forces its way in below the eaves of the house. Something must be wrong: this was April! Why, only yesterday it had been the dawn chorus which had welcomed me from my dreams of summer and the joy of running my dogs on perfect sheep around a velvet course.

Suddenly wide awake, I leapt from my warm cocoon and crossed hastily over to the window, where one glance was sufficient to put me in a panic. I reached for

Whose baby?

my clothes, my mind filled with anxiety for the ewes lambing high on the bleak and exposed hill-top, probably endeavouring to shelter in the lea of the dry stone dyke which, with the wind the way it was, could only act as a death trap to them.

Ten minutes later, warmly clad, with anxious collies Garry and Loch at my foot, I crossed the burn via a bridge which could only be reached by wading waist-deep through an enormous snow drift. Then the dogs and I scaled up the steep incline, the hazardous and most arduous part, which took us some two hours. Finally, with me near exhaustion, we reached the ewes. Those which had not lambed were persuaded, slowly and tediously, to go downhill into the safety of a small field near the house.

Although we were by now frozen and soaked to the skin, a return journey was necessary to gather up the ewes with young lambs at foot. As we began to climb, the wind worked itself into a frenzy, forcing the snow down my neck and into my face and eyes, where it stung like red-hot needles. I wondered if, on days like these, there was anyone who gave a shepherd, or his dogs, a sympathetic thought and whether those who worked in warmth, shelter and comfort ever counted their blessings (not that I would change places with them mind you!).

At long last we reached the ewes and lambs. Several ewes were almost buried with only their ears sticking comically skyward as proof of their existence, their tiny lambs huddled beneath them as they snuggled in for warmth and protection from the storm. The dogs, with snow-clogged paws, and me, with numb and frozen fingers, set about digging them out.

There should have been 17 lambs and, although we searched everywhere, I could only find 16. It was not until a couple of days later that I made the sad discovery of my lost lamb, curled in a tight protective ball where he had slept a few yards away from his mother's side. Having dug the sheep out and gathered them together, the difficult problem lay in persuading them to travel. If only the

GWB helping me bring the sheep home.

ewes would lead on, the lambs would follow, but the ewes were anxious for their infants' safety and were afraid to allow them out of their sight. My only option was to gently take hold of each lamb and, as carefully as possible, throw it from one snow drift to the next.

Halfway home and completely exhausted, I had to rest. The dogs, their coats by now weighed down and their whiskers stiff with silver frost, snuggled up close as we tried, unsuccessfully, to draw a little warmth from each other's steaming bodies. Momentarily I closed my eyes and leaned back on what I took to be a bank of snow. Rudely awakened I gave a loud cry (scattering the poor dogs) and quickly leapt to my feet. Underneath the snow, its golden blooms peeping at me provocatively, was a large and prickly gorse bush!

After regaining my composure, I couldn't help but smile (if somewhat ruefully!). Such is life. Nothing it seems, is what it appears to be, the irony of it all! On we plodded, the going getting rougher, Garry suddenly gave a low growl through icicle-bedecked muzzle. In the distance, a lone figure, my husband, with fresh and frollicking dogs at his side, was slowly wending his way in our direction, having finished his own tasks on the steep Whitelaw. I was never more overjoyed to see anyone in my life. Gathering the remnants of my favourite old raincoat around me, its tattered hem flapping noisily in the wind, I seated myself in the snow (more carefully this time) to await my Sir Galahad patiently!

Everything was soon under control and an hour later ewes and lambs were happily nibbling and munching at sweet-scented hay within the sheltering walls of the old stone sheep stell. A quick glance at the first flock of ewes I had brought down some seven hours earlier showed me that only two out of the 90-odd that were left to lamb had given birth, and they and their babies were snug and safe within the centre of the rest of the huddled sheep which had obligingly pressed tightly in around them, forming an impenetrable shield against the fierce and raging storm.

The compliment

A number of years ago, when employed as a shepherdess lambing an 'in-bye' flock of 'Mules', 'Mashams' and 'Halfbreds', one of the nicest compliments I ever received was from the ewe which returned home to lamb. Each evening, for the sake of convenience, I would bring the 'in-lamb' ewes into a 9 acre area adjacent to the farmhouse, in full view of my bedroom window. Any ewe which gave birth to twins or triplets was immediately drafted into a cosy pen within the confines of the 'maternity wing' which was situated in the corner of the field.

Early each morning, after feasting on what was to them a delicious breakfast of oats, flaked maize, linseed cake and juicy locust bean (which, when given the opportunity, I hastily picked out a few for my own use!), the ewes were allowed access to 40 acres of grass next door. On the morning in question, after emptying their ample feed from my 'corn bag' into the long line of troughs, and waiting patiently whilst they trotted speedily back and forth with glazed expressions, hoping for more, I opened the gate, counting them as they drifted through. Some made for the water trough on the far side of the field, whilst others stopped to nibble nonchalantly at a turnip before wandering on to graze on the lush green grass in the hedge backs.

When all the ewes were through the gate, I left it open and set off homeward to feed the orphan lambs and eat a hearty breakfast myself. I then called in at the lambing shed to see what troubles awaited me there. On this particular morning I was pleasantly surprised for, within the largest pen (normally reserved for triplets only!) in all her glory, contentedly nibbling alternately on a juicy turnip and the sweetest of hay there stood a large dewy-eyed, mottle-faced Masham ewe! Closer inspection revealed, still wet from her womb, but none the less adorable, triplet lambs nudging at her woolly flanks on wobbly but sturdy limbs. Considering that a ewe's natural instinct tells her to give birth in a quiet and isolated place, I believe that she paid me the greatest of compliments.

The foxes
As my bedroom window conveniently overlooked the lambing field, on rising, my first reaction was to glance outside. On one occasion, I could hardly believe my eyes for, down at the far dyke, there were two sleek foxes (presumably a dog and a vixen) plaguing a ewe with a newborn pair of lambs. One fox was trying to hold her attention by snapping at her nose, whilst its mate tried to catch one of the lambs. I could not for the life of me think what to do for the best and, for a moment, I froze. I knew that it would take me at least five minutes to pull on my boots and race down the field. By that time, it would more than likely be too late. My two dogs were housed together in a large kennel beside the garden gate and so I did the only thing possible. I rushed down the stairs and, even before they had time to greet me, sent them speedily down the field. Knowing that it would take them only a minute to reach the scene, I hurried inside to dress. By the time I got to the sheep it was difficult to believe that there had been any foxes in the vicinity. The dogs were already returning from a fruitless chase which had taken them across two fields. With pink tongues lolling and heaving flanks, they raced to my side to wish me a belated 'Good morning', whilst the ewe, her near tragic experience completely forgotten, grazed proudly alongside her now sleeping babies.

These same two foxes were to return to the scene regularly until the vixen produced her four cubs in a large earth a mile away. After that, in the weeks that followed, the dog fox was to return alone doing untold damage by splitting 6 pairs of lambs and mauling a 'sway back' lamb. This particular fox's tactics were to catch the lambs by their tails before removing the head which he dutifully carried home to his wife, after first covering the remaining carcass over with grass. Later he would return, eat most of it and take the remainder home. An old and well respected shepherd eventually came to my rescue, advising that I put a smearing of melted Stockholme tar on the tip of the tail and back of the neck of twin lambs. Single lambs, he said, were well taken care of by their doting mothers and triplets, ever hungry, clung closely to their mother's side. It was twins which were the temptation. After tarring I can honestly say that in the following ten years I have never lost another lamb to Mr Fox. (Mind you, if a still-born lamb became available it was left in a convenient corner so as to ensure that the fox's hunger did not lead him into mischief!)

A proud mother guarding her offspring.

In the late evening I would often sit downwind watching the cubs at play. One evening I was spotted by the vixen as she returned with a plump chicken from a neighbouring farm. In the ensuing commotion the bird was dropped as both vixen and cubs disappeared into the spacious earth. Within seconds, out popped the largest of the cubs, who promptly retrieved 'the supper' and quickly vanished with it back down the hole!

I have seen a replete fox actually playing with a live lamb, before going about his business, leaving it unharmed. That same fox, should he find his way into the poultry house would, in the ensuing commotion, be more than likely to slaughter each and every occupant. I cannot find it in my heart to hate the fox. Dogs and humans are equally capable of his crimes, however, I realise that he must be controlled. It is the method of the control that is the problem. Unfortunately no one has come forward with a satisfactory humane method of dealing with him. With rabies an ever increasing threat, the day could come when both the Government and the landowners will need to act with great haste and then the fox, like his more noble brother, the wolf, will be no more. The wolf has not been missed. The passing of the fox will be mourned only by the hunting fraternity.

Capturing a wild beast!

On cold wet nights when the wind howled, my old alarm clock would be set for 3 am just in case a weak lamb should decide to come into the world, and perish before the dawn. On one such night I awoke at 2.30 am to the sound of heavy rain spattering against the windowpane. Once outside, I made in the direction of the furthest corner—a favourite spot for ewes to give birth. My torch's piercing beam picked out an old draft ewe, a wild and woolly Blackface with a broken horn. We had purchased her the previous autumn at the sales, because she was too old for the hill. Startled, she turned quickly away from the light and, as she did so, I noticed that she was in difficulty for, swinging like a pendulum from her rear hung a lamb's head, huge and somehow grotesque in the shadows. Its forelegs were bent backwards trapped within her. One wrong move on my part and she would have undoubtedly vanished into the darkness. For exactly the same reason I was reluctant to try and catch her with my dog.

From the next field came the plaintive cry of a young lamb, wondering at the torch's beam. The Blackface swung round quickly in the direction from where the cry came, anxiously calling out a reply. It gave me an idea. I whistled quietly, bringing my dog to where she could see him and then I bleated loudly, imitating the sound of a new-born lamb. The ruse worked; my ewe appeared completely fooled. In her confused state she firmly believed that she had already given birth, and furthermore, the bedraggled apparition (namely yours truly!) which was now confronting her had stolen her baby! The dog, of course, 'she' was totally convinced, was about to devour it! She emitted a loud and furious snort of anger, at the same time making towards me. Coward that I am I took to my heels and fled rapidly in the direction of the lambing shed, remembering to keep up a continuous bleating until I arrived, somewhat breathlessly, at my destination with 'her ladyship' hot on my trail. Hurriedly I fumbled for the light switch and switched it on, to be greeted joyously by a loud but extremely effective chorus of orphans basking below the glowing warmth of the infra-red lamp.

The old ewe answered them frantically and rushed passed me, charging into a vacant pen. Breathing a sigh of relief, I hurriedly closed the gate behind her. I had captured the wild beast! Within seconds, I had her tipped on to her right side and was probing carefully inside her. I slid my right hand down over the lamb's shoulder, wincing as the ewe's pelvic bones tightened on my knuckles as she endeavoured to strain. My fingers slid easily behind each knee—thank goodness she was old and roomy and so I soon had the feet positioned correctly. I glanced at the lamb's head—his already protruding horn buds told me that he was a male. The ewe gave a mighty heave, at the same moment I pulled downwards, clasping the lamb's neck in my right hand and his slippery legs in my left. Suddenly there was a whoosh and out he slid, steaming, glistening and wet, his flanks white as the driven snow, bespotted with crimson blood as the lifeline that joined him with his mother was severed forever. I quickly cleared his nostrils and mouth, at the same time rubbing his back, immitating the stimulating licking of his mother's tongue. He obligingly coughed, spluttered and shook his head. I smiled, relieved that he would live, and gently placed him at his mother's nose. She scrambled to her feet, all the while making low mutterings of encouragement as she cleaned the mucus from his warm and gently heaving body. I climbed the gate of the pen and watched them for a brief moment, before creeping quietly away, leaving them to get to know each other and form the vital bond which would last all through the spring and following summer.

Above *Duties of a shepherdess and her dog. Caring for the lambs.* **Below** *The old ewe with Tiny, her adopted lamb, clad in her dead offspring's skin.*

The old ewe

I was given an elderly ewe at a local sale, which no one wanted, presumably due to the fact that most of her teeth were absent! Lambing time arrived and, with little effort, she produced a large single lamb. I had hoped for triplets or even twins as she was enormous. Disappointed, I turned her away to run with the other singles. Two days later, surprise! surprise! out popped a second lamb—a delicate, miniature creature, perfect in every way. I congratulated the old ewe and her sagging jaw appeared to give me a toothless grin. She certainly had the last laugh though for, on turning her up, I discovered that she only had one teat and subsequently could only feed her firstborn! Fortunately I had a bereft ewe with an udder just brimming with milk who happily adopted Tiny who, within five minutes, was placed at her side, clad securely in her dead offspring's skin.

The blue-faced Leicester

A few years before I was given the toothless ewe, I made the acquaintance of a large fat Blue-faced Leicester ewe which had been purchased surprisingly cheaply—presumably a non-breeder. (We found out at a later date that she had failed to produce lambs when mated to a ram of her own kind.) She was 4½ years old when she came into our possession and we immediately put her to a Suffolk ram. Obviously not 'colour prejudiced' she went on to produce 12 bonny cross-bred lambs within the space of four years, biting off all of their tails to the 'correct' length within a few minutes of birth!

My first hill lambing

During my first lambing on the hill, I turned away a ewe with a strong, week-old lamb at foot, leaving them just through the gate so they could return to the terrain where the ewe herself had been reared. They had to cross a burn first. Unfortunately, greedy for the tender young grasses which had grown on its banks in the absence of the in-lamb ewes, both the ewe and the lamb wandered upstream. Finally the ewe, by now replete, chose a spot where the water was at its deepest and leapt across, presuming that her lamb would follow, but he was afraid and after dithering on the edge, flatly refused.

Most of that day the rain fell heavily and I was kept busy herding the newborn lambs to shelter. When evening came the sun was shining once more as I set off for my last rounds of the day. As I neared the gate leading out of the lambing fields and on to the hill, I thought I heard the hoarse cries of a lamb and hurried through the gate and upstream just in time to witness the ewe leap into the icy, fast-flowing water, where she stood motionless whilst her (by this time extremely hungry and thirsty) lamb scrambled over her woolly back and, with a hop, skip and a jump, ran safely up the opposite bank.

Since that day I am particular to check that any ewes and lambs which are put out to the hill have safely crossed over the burn before I return to my duties elsewhere. All living creatures are far more intelligent than we ever give them credit for. They have so much to show us, if only we would spare the time, so much to teach us, if only we cared to learn and such a lot to tell us, if only we would pause to listen.

Rounding up the flock.

Man constantly interferes in their lifestyle, doing what 'he' considers is for the best and using the economy as his excuse. Little does he realise that by interfering with nature he does more harm than good in the long run. Left, where possible, to organise its own life, the average creature will make a much better job than his human counterpart.

The more hours I spend in the tranquil company of animals, the more I realise how relatively slender is the line which divides us. Animals have taught me that if I am prepared to listen with my eyes, rather than my ears, their language is easy to understand. For their message to us is a simple one; their requirements are modest. They seek only peace, gentle care and, most of all, a greater understanding of their ways.

Chapter 15

My sheepdog trials experiences

One Man and His Dog, 1981

In the spring of 1981 an exciting invitation arrived at Swindon from Ian Smith, the producer of BBC television's *One Man and His Dog*. GWB, Jed and Trim were to appear, running in both brace and singles in the televised sheep dog trial. It was decided that Jed would run for Scotland in the singles. Although of the two bitches she was the most difficult to handle, she could be relied upon to work with intelligence and determination in a sticky situation. Both bitches were at that time 10½ years old. Apart from Trim's short-sightedness and Jed's convenient deafness, they were, and are, remarkably fit for their advanced years.

It was a gloriously sunny day in the latter part of July when Ian Smith, accompanied by Margaret, his cheerful smiling production assistant, complete with camera crew, drove up the winding valley road to film us and the dogs on our hilly home ground. For weeks before this special occasion, I had an ever-anxious mother priming me with tactful telephoned suggestions as to 'my apparel' for the great day. Geoff, Junior, was much more to the point and completely lacking in tact! I believe he thought I was quite likely to appear in front of the camera in my lambing garb, or Worzel Gummidge outfit as he describes it!

Above *Prancing around in a blue velvet skirt.* **Below** *Geoff Jr didn't get off lightly either. Geoff Jr at a later date with renamed pony!*

Jed, Geoff Jr and Tess, the reserve.

GWB wasted no time beating out the bush. (The bush in question being my wind blown mop of hair!) 'You really must do something with it,' was the daily plea. Finally (and providing I was allowed to wear my 'wellies'), to placate the family I condescended to 'prance around' the Whitelaw hill in my best blue velvet skirt, and matching blouse, sporting wellies of the same hue! The offending 'mop' of hair, I was not prepared to do anything about other than wash it and brush it with extra vigour, however I really can recommend a certain brand of dog shampoo!

Little Lucy had just turned 10 weeks old when filmed by the BBC working Cheviot ewes and lambs up the steep face of the Whitelaw. On that day it was definitely a case of 'third take lucky'. She proved to be in one of her unco-operative moods. As I carried her from her kennel I could not fail but notice the mischievous glint in her rapidly changing baby blue eyes, and wondered what form it would take.

We did not have long to wait before finding out! I gingerly set her down a few yards from the sheep. She began to drive them prettily first across and then down the steep incline. I was just about to whisper 'show off', when all of a sudden, taking us all by surprise, the naughty little minx took off! What followed closely resembled a cavalry charge with sheep and men stampeding in all directions.

Once again the same carry-on occurred on 'Take Two'. By 'Three' I was beginning to wonder if it had been a good idea to attempt making a 'star' out of our little Lucy after all. I gathered the quivering bundle up into my arms at the same time whispering a final plea into her extremely alert ear, already pricked with anticipation. She finally got the message, either that or the game was

Below right *Jed, Trim and GWB. The favourite team on another occasion.*

becoming a bore. 'Take Three' was not only successful but in my opinion, rather touching too. The ever-patient Ian laughingly remarked, as he breathed a sigh of relief, that all good performers were entitled to be a little temperamental. (I swear Lucy gave me a wink as I popped her into her kennel.)

GWB Junior did not get off so lightly. Dinah, his elderly but lovable pony, embarrassed him and disgraced herself by frequently and loudly breaking wind at the most inappropriate moments. Since her television debut she has been aptly rechristened Dynamite!

The actual 'heats' and 'final' of the *One Man and His Dog* programme were to be filmed in September in the beautiful countryside of North Wales. The 14th of that month saw us driving down to Bala after depositing a carload of dogs with kind friends. (What would we do without them?)

We travelled quickly along the motorway that evening, the rain pelting ferociously against the windscreen, with the wipers dancing frantically back and forth, causing rivulets of water in their vain attempt to clear our view. I was beginning to feel sleepy, hypnotised by their constant movement, when, out of the darkness, there came a still small childish voice. 'Mum,' it said, 'are Tess and

Garry supposed to be getting married?' I looked round in surprise, shining my torch on the occupants in the rear of the car and, sure enough, there was Garry doing what in his case occasionally comes naturally, with Tess, (the reserve, who had unexpectedly come into Season), watched closely by a somewhat surprised and curious audience consisting of one small boy, Jed and Trim. We removed Garry at the first convenient moment!

Finally, after a long and somewhat wearisome journey we arrived at our destination and booked in at the same comfortable hotel that we had stayed in the previous summer when the International Sheep Dog Trials were held at Bala. We arose early the following morning and, after eating an enormous and delicious breakfast (with plenty of leftovers given to the dogs as an added treat!), we set off in the direction of Rhiwlas Farm, where the trial was to take place. There we met fellow Scottish team mates Tom Watson (full of fun as usual) with his bonny International winner Jen and Stuart Davidson along with his handsome bare-skinned National Champion, Ben.

As the Scottish team were not competing until the Thursday, and this being only Tuesday, we were able to relax, study the course and enjoy the beautiful scenery which included a view of Bala Lake. High on the hill we were in sight of where the very first sheep-dog trial was reputed to have taken place.

What I remember most about Rhiwlas is the peaceful undulating countryside with its lofty and majestic oak trees, scattered like friendly giants across its landscape, their gnarled trunks and leafy branches providing a welcoming shelter from the bitter wind that swept in off Bala Lake, bringing with it short sharp showers of freezing rain which were followed by brilliant sunshine and vast cloud shadows, travelling swiftly over the emerald earth.

I have not forgotten the kindness of the Welsh people, especially that of one Glyn Jones who skillfully set out sheep for the competitors. So typical of his kind, he invited us come evening, to follow him deep into the nearby heather-clad hills so that our dogs might stretch their legs on his flock of sweet little white-faced

Left *The presentation.*

Below *The trophy.*

Right *The Bowmont Valley. Setting for our trial in 1981.*

Below right *Jean Thompson and Wullie.*

mountain ewes. After being cramped up in the back of the car for many hours they were in need of exercise, and duly got more than they bargained for! The 'little Welsh sheep' proved agile and fleet of foot, requiring free, wide-running dogs with plenty of stamina to gather and hold them together. I believe that Jed and Trim, by the looks on their faces, were more than happy to return to the car!

The BBC were wonderfully efficient, considering what we trial enthusiasts are used to, and provided everything, including luxury loos. The last loo I had visited at a nearby trial was a tin can with a lid which carried a notice saying, 'Do not exceed ten stone'! A huge marquee had been erected for workers and competitors alike to have their mid-day meal within the shelter of its vast canvas walls. When the weather proved too chilly for the children, they too played and squabbled inside with pretty farm kittens which had taken refuge in the warmth whilst we spectators put on a brave face and turned up our coat collars against the cold wind.

I made good use of the weather, making it a justifiable excuse for the purchase of extra clothing in the shape of two extremely attactive and fashionable anoraks. One was in a delightful shade of pink, with a rosebud-patterned reversible lining, the other was also floral, in pastel shades of grey, blue and cream. (I just couldn't resist them!) After presenting GWB with the bill, and noting his rapid change of countenance, I had to do some quick thinking and pacified him with the excuse that my mother, his mother and a few million other mothers might see me on the telly, and how I was sure he would want me to look my best, etc, etc. (It worked!)

Our few days at Bala proved very rewarding. Seeing how the series was produced was a marvellous and exciting experience for me. It was rather like a giant jigsaw, each piece fitting into place. The efficiency has to be seen to be believed. By the end of the week, we really began to feel part of it all. And guess what—we pulled it off! Stuart Davidson and Ben from Dunoon won the singles class and dear old Jed and Trim, as their grand finale, won the brace. We travelled home exuberantly with our trophies to bonny Scotland. For determined Jed and gentle Trim, litter sisters, this must be the ultimate in their trialling careers and should they now retire from the scene, they could not, for me, go out on a sweeter note.

Ray Ollerenshaw, Chairman of the International Sheep Dog Society, made the presentation of the much coveted BBC trophies depicting a Border collie and a camera standing side by side, mounted on a base of polished wood. The tear I wiped from my eye on this occasion was one of pride.

The Bowmont Water Invitation Sheepdog Trial, 1982

In the early part of 1981 the idea of holding a sheepdog trial which would provide a stiff and practical test, over hill ground, was conceived by Major General David Lloyd Owen who then formed a Committee consisting of himself, Dick Fortune and my husband Geoff.

Permission to hold the trial was granted by His Grace The Duke of Roxburghe. The trial date was set for Sunday, October 10 1982. Invitations were sent out in the hope that at least 27 sheepdog handlers would accept. Twenty-six did, three of them ladies, which just goes to prove that the Committee held no chauvinistic

A relaxed and friendly occasion.

views. The ladies in question were Mrs Jean Thompson of Castle Douglas—a sweeter soul you'll never meet—with her devoted auld pensioner Wullie. (They were members of the Scottish team at Bala International in 1980.) Then there was Mrs J. Arneil of Braco, a consistent trialist with her pretty Gail, and myself with Garry.

Among the competitors no fewer than 17 had been in Scottish teams, six handlers had represented England and there were five Supreme Champions. Because of his many years of varied trials experience in this country and abroad, Dick Fortune of Edinburgh was asked to judge the trial.

It was decided that the most suitable site for the course was the south face of the Whitelaw. A great view was to be had for competitors and spectators alike, who were able to park their cars on the roadside below the course. Sheep folds and toilet facilities were permanent fixtures in the valley and it was arranged for drinks and sandwiches to be served from a converted railway carriage. Unfortunately the roof proved to be rather leaky but, luckily, nearby Sourhope came to our rescue with a large waterproof sheet which, when weighed down with old tyres, stopped the drips.

Preparations for the trial were begun two weeks in advance. Until the

The Split (Frank Moyes).

Bowmont Trial, I never realised that so much thought went into a course or that there were so many anxious moments. The place from where the 'lift' should commence was the biggest headache. For a start, it was so far to walk!

We tried 'lifting' sheep ourselves from three possible places on the hilltop, using dogs of various ages. We found that young dogs tended to get lost on their outruns. Experienced dogs would 'lift' sheep anywhere, but these were our dogs and they were familiar with the ground. Finally we decided on a place and positioned the white marker post across towards the outlet pen where no dog could fail to see his sheep. The pen itself, carefully camouflaged with a high sackcloth plus tree branches, was situated on the only suitable ground which was next to a wood.

GWB erected a well thought out course and a five-sided pen which consequently worked a treat. The five hurdles used bluffed those wiley little Cheviots into thinking that the pen was larger than it really was. We were worried as to how the sheep would behave on the course, so I spent the best part of a fortnight waking them first thing in the morning and quietly dogging them, still sleepy, downhill. As it turned out they behaved perfectly with the exception of poor Jack Thompson's 'Wild Bunch' and my 'Senile Delinquent'. Competitors were given two gimmers and four ewes. Providing six ewes for each handler would have meant using sheep off a different part of the hill—sheep which might not have been at all willing to participate on strange ground.

The competitors and officials (Frank Moyes).

Thankfully a saviour arrived a few days prior to the trial. This was a horse called Smokey Joe and his description is huge, magnificent and white! He's the kind of charger knights of old went galloping into battle astride. Smokey was a gift, and one must not look a gift horse too closely in the mouth. His temperament might be pure camel, but most important, he is absolutely harmless and incredibly wise. Without his help the sheep for the trial might never have been rounded up at all.

After a week of solid rain, a thick mist descended, enveloping the surrounding hills like a shroud. GWB was busy weaning calves so Garry, his son, Laddie, and I (riding Smokey Joe) set off to gather the hill. As I rode off I couldn't help but notice my washing, some 20 soggy garments in all, hanging limply and lifelessly on the line, where they had been all week.

We gathered the hill in two halves. Shutting the top half out of harm's way in a field on the roadside beside the house, the sheep being used for the trial were placed in a small field near the course. By the end of that wet and misty morning, I fully appreciated the many advantages of working dogs from high up on a horse. Apart from realising that nine times out of ten a dog working out of sight is using brains and initiative and therefore is best left alone, riding a horse is a helluva lot easier than walking.

It was still raining heavily the night before the trial and so I reconciled myself to the fact that, like it or lump it, I was just going to have to suffer the indignity of

Penning (Frank Moyes).

a line of wet washing on display. Sunday, October 10—trial day—dawned cloudy but thankfully the rain had ceased and, apart from the odd shower, the day kept fine. We stood and surveyed the course, it included a near enough 700-yard outrun over difficult terrain, rushes, drains, a stone-built stell and, thanks to the inclement weather, a water jump!

For me this was an historic and friendly occasion. Throughout the day there was great camaraderie and interest in what was taking place. The advantage in having fewer competitors at a trial is that they tend to socialise more; their interest rivetted by how the dogs would tackle the course rather than on who was going to win.

The only part of the trial not visible was the position from where the dogs actually performed their 'lift'. For practical reasons the sheep were released on the skyline. The fact that each 'packet' of sheep came away steadily and 'on course' could mean that too much emphasis is placed on where and how a dog actually performs the 'lift'.

Perhaps, if left to their own devices, the dogs might teach us a thing or two. I have often thought the distraction of men and dogs at the letting out post can, and does, play havoc with the latter part of the dogs outrun and 'lift', resulting in some judges having a field day with their pencil! So, 'let the dog see the rabbit'!

An added highlight of the Bowmont Trial was the presence of His Grace, The

Above *Glen in his winning run.*

Below *Leadburn Nell, Garry's mate and mother of five nicely marked litters.*

Duke of Roxburghe, who, as it turned out, had the pleasure of presenting the first prize to a shepherd in his employ, for the first prize went to Geoff and his rough-coated Glen, the son of Leadburn Nell and Garry.

The Scottish National, 1982

The 1982 Scottish National Sheep Dog Trial held at Strathaven (Straven), Lanarkshire will go down as one of the best. It was certainly ideal from both the competitors' and spectators' points of view. Held on a gradual slope and clearly visible to all, the course was well situated with even the smallest of details taken into consideration by the local committee.

Over the three days, the Blackface sheep were generally well behaved, although on occasions a competitor was presented with a 'three' and a 'two' to nurse around the course, and consequently it was the most dominant and masterful dogs which came to the fore. The weather proved near perfect although it was perhaps a little warm for sheep and dogs alike, on the Thursday. Thankfully there was no reappearance of last year's 'Iron Lady of Glamis' cattle pen. Sheep were placed beautifully at the post throughout the three days by 'one man and his dog'!

I set off for Straven with little thought in my mind of 'making' the Scottish team. Garry had always given of his best at previous Nationals but this, after all, was our fifth and we had failed at the previous four attempts. I even wore a dress, something I had not even contemplated before. This particular garment was flowery and flowing. It was also, much to my embarrassment, prone to blowing up in the slightest breeze.

Anyway, our turn arrived and out we went—number 23 on the Saturday. The sheep, if anything, were slightly cheekier I noticed, and rather more suitable to Garry's somewhat determined tactics. I stepped up on to the stand provided for the 'not so tall', like myself. I am firmly convinced that at a sheepdog trial tall people have an unfair advantage. For a start they can look right into the obstacles and are all arms and legs at the pen and shed.

Our sheep were at the post, my heart sank as one of them turned and walked away from the others. Garry was already away right-handed and, without any commands, he gently gathered up the offending ewe and put her quietly to the others. Apart from a few deviations around the course we managed somehow to get all the sheep through the gates—quite a pleasant surprise. I heaved a sigh of relief and stepped down into the shedding ring.

However, the shedding proved to be somewhat awkward, to say the least. My off-white anorak and billowing skirts scared the life out of those suspicious Blackies. Every time I made any attempt to advance they retreated to the very edge of the sawdust shedding ring. Finally, in sheer desperation, I 'barged' Garry in as a last resort as time was getting on. Luckily the rough tactics paid off as Garry cut the two collarless sheep from the others. I walked over to the pen leaving 'his nibs' to rejoin the sheep and fetch them after me. I don't know what he threatened them with but, greatly to my surprise, they bolted in to escape it.

When it came to the shedding of our red-collared single it just had to be the same tactics as before, and thankfully they paid off and we were home and dry. Our 1982 National 'run' was over and our score from judges Andrew Ferguson

and Willie McMillan was 185, the second highest for the day after John Angus MacLeod's stylish and impressive performance with his Ben. (Ben impressive and John stylish!) Overall, Garry and I came into the 12th place on the team for Scotland.

I was living in a dream, two days later the dream became reality when, finally, I realised that I had achieved my second greatest ambition in life. I climbed with my devoted Garry to the highest peak on the Whitelaw from where you can look down on the loveliest view imaginable and, after a careful glance around to make sure no one was about, I shrieked 'Yipeeee' at the top of my voice, startling the sheep and a distant family of crows as the echo reverberated around the lonely hills. Garry looked up in astonishment as I collapsed laughing beside him. He licked my nose and I gave him a great big hug of gratitude.

In the weeks that followed, apart from practising our shedding 'International style' and one 'double lift' away from home, I deliberately gave Garry no special training. He was used to long gathers on the hill and was often put back for different cuts of sheep in his everyday work. I went off to the Blair Atholl Inter-

Left *He licked my nose* (Ian Hossack). **Below** *A critical spectator* (Frank Moyes).

national admittedly extremely nervous, but with a great deal of faith in my little partner. Should we be fortunate enough to qualify for the final day?

The International, 1982

The atmosphere and the excitement of my first opportunity to compete at an International Sheep Dog Trial will remain in my memory for ever. Held in the lovely undulating parkland surrounding the fairy-tale castle home of His Grace, the Duke of Atholl, at Blair Atholl in Perthshire, on a long, low-lying field, the event was judged by Peter Hetherington from Scotland, G.W. Jones from Wales and Harry Huddleston from England.

Garry ran in the Qualifying Trials on the Thursday. True to character he never put a paw wrong, which is more than I could say for myself for, thanks to me, we missed all our gates and ran out of time at the final shed. Unfortunately, the Blackface sheep were poised for flight and were far too light for a dog with Garry's presence. It took a good handler to steer them through the hurdles at the speed they were travelling.

The International Course for the Championship.

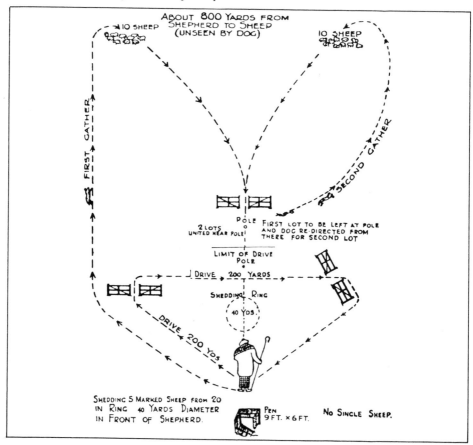

Although I failed to get my dog through to the final day, the experience was well worth the effort, and I did not come away empty-handed. The French contingency brought with them an enormous ornamental glass bottle containing a lovely perfume which was duly presented to me from a dear friend's Wellington boot where she had placed it for safety's sake. Now on special occasions, when I use the perfume, I am reminded of the excitement of the Blair Atholl International, but mostly I am reminded of the kindness of the sheepdog enthusiasts who nursed it all the way from France. My proudest possession to date was presented to me in the grandstand by Thomson McKnight. It is a gold enamelled badge, depicting Jock Richardson's Wiston Cap and presented every year to each member of the teams.

After Thursday's run we travelled back home to feed and exercise our other dogs so that we could return to Blair Atholl to watch the Supreme Championship on the Saturday. On arriving home we discovered that Trim, within a stone's throw of her 12th year, had given birth to perfectly marked twins. The newcomers were christened on the spot—Whitelaw Trim II and Whitelaw Jed II.

The new arrivals. Whitelaw Trim II and Whitelaw Jed II.

The Saturday dawned warm and sunny and, after an early start, we were soon comfortably installed high in the grandstand at Blair. The Blackface sheep were noticeably more manageable that day and I would have given almost anything for the chance to attempt the 'double lift' with Garry. Instead I could only close my eyes and dream. (It's often less disappointing that way.) At the end of the day Welsh team member E. Wyn Edwards and his handsome rough-coated Bill became the stars of Blair Atholl and this was their second Supreme Championship in succession.

The International—a bird's eye view

On reflection my 'female' view is that every competitor who qualifies at a National to run at an International should be allowed to run their dogs in the Supreme Championship. As it stands it largely depends on the sheep's behaviour, Lady Luck and the skill of the handler whether or not one gets through the qualifying trial. A much better way of ensuring that everybody is given a fair chance would be to take the ten highest pointed dogs at each National and give them all the opportunity to attempt the double lift at the International. While we are on the subject, why not have a mini double lift with each competitor limited to one dog of the handler's choice at National Sheepdog Trials? It would certainly make for a more practical and entertaining National by encouraging the skills of the genuine shepherd and his dog.

Another point well worth a mention, according to those concerned, is the location. Competitors are fully aware that it isn't only the field that has to be taken into consideration, although, along with the sheep, it should be the prime concern of those involved in its choice. According to experienced campaigners, large, flat, pancake-type fields are not suitable places to hold an International. I list the reasons given:

1 Distances are extremely difficult to judge on the flat.

2 From the dog's point of view, hearing can be and often is a problem.

3 On a large flat field, sheep tend to gallop, especially the athletic hill types.

4 Unless both the judges and the spectators are placed high up in the grandstand, viewing a dog's behaviour when tucked in behind his sheep on a flat field is virtually impossible.

So why, with three years for each country to find a field or site, are International Sheepdog Trials regularly staged on the flat?

The 1981 International, staged on a hill-type course at Nord View Farm, Armathwait in Cumbria, was a fair and practical test of both dogs and handlers and, from the spectators' point of view, a joy to behold. Please let there be more courses of this type. They are excellent entertainment value for the public. Most important, they provide a challenge to the collies and competitors for, after all, there would not be a trial without them.

One Man and His Dog, 1982

Monday, September 13 1982 once more saw us on the road, this time heading in the direction of Fairlie, situated just outside Largs on the west coast of Scotland. In the back of the car was a precious cargo—Geoff Junior, sleeping soundly, using Garry as his pillow and a very pregnant Jed to warm his feet.

The letter, marked 'From the BBC' had arrived way back in April. I naturally thought that it was for GWB until, on examining it more closely I saw that it was addressed to me. My pulse began to quicken, could it be? No, surely not. I had only just returned from the lambing field so I quickly washed my hands before tearing open the long white envelope and sure enough, there before my eyes, it was! An invitation from producer, Ian Smith, to compete in the *One Man And His Dog* television Sheepdog Trials.

We were filmed 'at home' on a tempestuous summer's day (I having been discovered clad in my best dress furiously scrubbing the kitchen floor, in readiness for my visitors!) Out on the hill Garry and I were caught for several minutes in a torrential downpour until Ian prompted a certain gentleman to come to my rescue. I was bundled hastily into a vehicle already brimming over with people and equipment whilst poor Garry lay outside getting wetter and wetter, guarding a small flock of impatient Cheviots.

Presently the sunshine returned and Garry was speedily rubbed dry with a couple of grubby handkerchiefs discovered in the depths of a raincoat pocket, whilst I endeavoured to drag wet fingers through my dishevelled hair. With the help of a warm breeze, repairs were quickly completed and the necessary filming soon accomplished, with Geoff Junior and his pet lame duck being immortalised on film forever. Ian, Margaret (his production assistant) and the film crew departed once more carrying with them, we hope, fond memories of Swindon and Bowmont Water.

So here we were on a bright Monday morning in September 1982, heading westwards to have a look at the trial course in readiness for the following day when the Scottish heat of *One Man and His Dog* was due to be filmed. On arrival we were greeted by Alistair Cutter from Stow, his wife, Sandra, and Alistair's black, white and tan 'classy' bitch Midge who, along with John Templeton and his bare-skinned regular prize winner Roy, completed the Scottish team.

At first sight, the course filled me with trepidation, however, after walking over it no less than three times, the tricky layout began to grow on me. It was certainly an interesting course, quite exciting. In fact it was a real challenge and I was determined to enjoy every moment of the following five days' competition. I had always told myself that the day I started to take trialling seriously, as far as winning was concerned, was the day I would stop. To me, working a dog is fun with a capital 'F', especially when the dog has initiative.

The course was situated high on a rock-strewn slope, with marvellous views to the left of the sparkling silver sea and the islands of Cumbrae with the shrouded peaks of Arran peeping over the hills in the far distance. A shady forest flanked the course on its right-hand side, sheltering it from the blustery wind. Across the centre a further row of trees and leafy thorns hid a deep ditch.

After watching a 'trial run' put on by a local shepherd and his young dog for the benefit of the cameras, we booked in at a comfortable hotel on the sea front along with Alistair and Sandra. Our room was situated on the ground floor conveniently looking out on the car park, so we were able to keep a careful watch on Jed in the back of the car where she was expected to give birth to her puppies.

The morning dawned grey but thankfully dry and after a large delicious breakfast we made our way back to the location where, along with the judges, Mervyn Williams and Thomson McKnight, we were briefed by Eric Halsall and shown a plan of the course inside a caravan parked in the narrow lane overlooking the set. Our questions answered, 'the draw' was made to establish who would run their dog first.

Woman's intuition told me that it would be me, and I was right!! It was. After lunch I sallied forth, microphone carefully out of sight, bedecked in a new lacey turquoise blouse and my old blue jeans. There I stood with Garry calmly by my side gazing up at me, earnestly watching for his cue. At long last it came in the form of a wave from stage manager Keith Hatton and we stepped out to the stone marker at the edge of the shedding ring.

I had decided to send Garry away to the left, just in case the seven Blackface sheep fancied a swim and decided to make a bolt downhill in the direction of the sea! Garry ran out well, silhouetted against the sea, the islands and the sky, and quickly arrived directly behind his sheep. They were absolute angels and 'lifted' quietly, happy to stay 'on line'. They went through the first set of hurdles then down into the dip in the centre of the course.

Anxiously I awaited their re-appearance, standing on tiptoe and sighing with relief as they came into view, slowly crossing the shallow water in the ditch. The 'drive' was to the left, down a steep embankment. However, we were allowed to walk behind our sheep, halting on a large glistening rock aptly christened, for this occasion, 'The Shepherd's Rock'. From there the dog was on his own, he drove his charges further down the slope and through a gap in the hedge.

Once I saw that Garry and his sheep were safely through the trees and over the ditch, I was allowed to make my way back to the shedding ring, at the same time peering through the thicket in order to complete the cross-drive. Suddenly, I froze in my tracks as, through a gap in the hedge, I witnessed my seven 'angels' calmly walking in single file on the wrong side of a white marker post with Garry by now thoroughly enjoying himself following quietly in their wake. I groaned inwardly, I had completely forgotten the existence of that post. There was nothing I could do but try for a 'good finish'.

The sheep were turned around the hedge at the end of the cross-drive and brought through two hurdles before arriving at the sawdust marked shedding ring. The ring was small and it took a few minutes to position the seven sheep so that the red-collared 'gimmer' could be cleanly shed without her 'companions' leaving the ring, and then on to the pen where they obligingly walked into the mouth and stood leaning heavily against the open gate. Garry stared at them whilst I waggled the rope and gently tapped the ground with my stick remembering the other occasion when I had done the same with a borrowed stick. Unknown to me, it was riddled with woodworm with the result that it snapped and I was left looking rather silly with only the horn handle to assist me.

I moved Garry fractionally to his left, for a moment my heart was in my mouth as one sheep momentarily made a tiny dart for freedom. Quick as a flash, Garry neatly turned her back and the seven sheep walked meekly into the pen. I quickly closed the gate, called Garry to my side and awaited Keith Hatton's beckoning which was the signal that the cameras had ceased to roll.

The Scottish heat was won by John Templeton with Roy. Garry and I were second and Alistair and Midge a good third.

Ian Smith would have liked to record the birth of Jed's puppy on film had she delivered it on cue. Sadly Little Spider, who was born a few days after we returned home, lived only for three days. Jed was just too old and far too stout for an easy birth.

One woman with her flock, 1982 (Ian Hossack).

Chapter 16

Omega

Until recently, springtime was my favourite season of the year, with summer running a close second. Early spring wears a special fragrance all of her own, a warm balmy aroma which drives out the harsh rigidity of winter, leaving one with a relaxed glad-to-be-alive feeling. I still go dewy-eyed with wonder at the sweet sound of the dawn chorus and the sight of the season's first lamb. Golden primroses, bathed in sunlight, growing wild on a mossy bank will always leave me enthralled.

The feel of warm spring rain refreshes my skin and the breeze carries the delicious scent of damp earth to my nostrils. At dawn the perfume of lush young grasses and vigorous spruce intermingles deliciously. Flimsy cobwebs, soaked in dew, shimmer and sparkle to my delight.

Perhaps, like everything else, our preference for the seasons alters with time for now I find myself sighing with relief as autumn puts in an appearance. It's a time to meditate, a time to relax, a time of fleeting Indian summers and cosy evenings by a log fire. It's a time of new beginnings when ewes yearn no more for their departed offspring, turning their minds to other things as they graze contentedly on the smooth, close-cropped terrain. It's a time when swallows bid farewell. There's no rushing away at the crack of dawn, with bleary eyes to distant sheepdog trials. No anxiety about the creatures left at home. I wonder at the sight of bright rowan berries and vivid hips and haws.

On the hill, sleek cattle with glossy, lick-marked coats follow me inquisitively through rustling fronds of dying bracken. Bare stubble patiently awaits the plough. I listen expectantly for the music of clanking stones on metal, the sound of sighing earth as it is deftly turned dark and inviting, exposed to the ever circling birds.

Cheviot lambs graze the emerald fogs, pausing only to bleat a muffled welcome, their mouths filled with luscious grasses, their dams long since forgotten. The trees are gracefully adorned in mantles of gold, bronze, yellow and faded greens. There's a sharpness in the air which invigorates, and a reassurance in the knowledge that all is safely gathered in. I bid you welcome, peaceful lovely autumn.

Epilogue

I hope that history will repeat itself for, slumbering peacefully in their kennels, are a new generation of Border collie puppies for GWB. Jan Mark Two is out of a sister of the original fast-flanking 'bionic bitch'. Sisters, Whitelaw Jed and Trim II lie dreaming of the day that they will run as one.

And for me there's the white-browed 'chiel' from the latest union of Garry and sweet Leadburn Nell.

Cosy evenings by the fire.

Glossary

Bare skinned Short haired.

Bomb sheep, to Bring in sheep too fast.

Brace Handling two dogs at once.

Cattle crush Metal crate for holding cattle.

Cow-hocked When a dog's hocks turn inwards, instead of being straight.

Cross-driving Driving sheep across the field in front of the handler and in a straight line.

Cross his course The perfect outrun is pear-shaped. The dog that runs between his handler and the sheep has 'crossed his course'.

Cut of sheep A flock of sheep hefted or running on a certain area of land.

Dew claw An extra, unnecessary claw on the inside of the hind leg.

Draft ewe A ewe too old for the hill which then is sold for breeding to a lowland farmer.

Drive, to When the dog takes the sheep away from the handler, usually on a triangular course through two sets of hurdles.

Easily worn When the sheep give way to the dog.

Easy dog A dog that is easily handled.

'Eye' The hypnotic stare, with which the dog fixes the sheep.

Fank Sheep pens.

Flank freely When the dog runs round the sheep without stopping or without any commands.

Fetch, to The bringing of the sheep by the dog to the handler. In an average Sheep Dog Trial this must be in as straight a line as possible through two hurdles.

Flanking When the dog runs round the sheep, either to the right or the left.

Fogs Clean grass or grazing which has grown up after the hay has been cut. Used to fatten lambs or for getting ewes into good condition before mating.

Gimmer Young female sheep.

Grass chewer A dog that pulls up mouthfuls of grass and spits them out while it works for no apparent reason.

Gripping When a dog bites the sheep. At a trial a dog is disqualified if he does this unprovoked.

Halfbreds Sheep from a Border Leicester sire and a Cheviot dam.

Haugh The valley bottom.

Heavy sheep Sheep that are difficult to move, usually because the dog lacks determination.

Heavy side of sheep, the The direction in which the sheep are trying to escape.

Hoggs Young sheep between a year and 18 months old.

In-bye Lowland fields.

Kep When the dog runs round and heads the sheep.

Lift, the The lift occurs at the end of the outrun and before the fetch. It is of great importance as it is the first contact the dog has with the sheep.

Light sheep Sheep that are easily moved and upset.

Line Ability in the dog to bring in the sheep in a straight line without any commands from handler.

Mashams Sheep with Teeswater sire and a Swaledale dam.

Mastitis An inflammation of the udder of a ewe caused by bacteria. Most common in cold weather.

Mindrum blood Prefix used by Bob Fraser for his famous breed of collies. Mindrum was the name of the Northumberland farm where Bob worked as shepherd.

Mules Cross-bred sheep. Sire usually a Blue Faced Leicester and the dam a Swaledale.

Out-cross Bringing in an unrelated and new strain of collie.

Outrun The course taken by the dog when gathering the sheep. At a trial the ideal outrun is pear-shaped with the dog coming in well behind the sheep.

Overflank When the dog runs past the sheep.

Pen A small enclosure usually constructed of hurdles, into which the dog and handler put the sheep without touching them.

Recast When the dog is sent back by the handler for more sheep.

Rookin Whites An old and famous strain of white collies which were bred at Rooking near Rochester, Northumberland.

Self-coloured A dog with a coat of one colour.

Shed, the When one or more sheep are separated from the flock.

Sheep nets Barrier or light-weight fence comprised of wooden posts and sheep netting. An easily moved sheep enclosure.

Shushing Noise to encourage dogs to keep going at speed until he gets to the back of his sheep.

Sourhope A neighbouring Research hill farm to Swindon.

Stell Stone built circular sheep enclosure.

Stop, a good A pronounced dip where the dog's nose joins his brow.

Stotting sheep A sheep that jumps kangaroo-style at great speed on all four feet.

Straight gather, a When the dog casts out for the sheep not in the ideal pear-shaped outrun but runs straight up the middle of the field.

Strong dog, a A dog that shows above average determination. Can sometimes be difficult to handle.

Sway-back lamb Lamb with a copper deficiency which sways about.

Tacketty boots Boots with rows of tacketts or metal studs on the soles to prevent the wearer from slipping.

Take time Command to dog to go slowly so as not to frighten the sheep.

Trods Paths made by the sheep through continual usage.

Tup A ram.

Undershot When the lower jaw is shorter than the upper one.

Wall eye Light blue eye.

Warms up When a dog develops more interest in his work.

Weak dog One that lacks determination and has difficulty in moving sheep unless they are timid.

Work on his feet Some dogs are prone to lie down while working. This method is less effective when working stubborn sheep.

Index